Power System Fundamentals with PowerFactory

Navid Bayati · Mohammed Ali Khan

Power System Fundamentals with PowerFactory

Springer

Navid Bayati
Center for Industrial Electronics, Institute of Mechanical and Electrical Engineering
University of Southern Denmark
Sønderborg, Denmark

Mohammed Ali Khan
Center for Industrial Electronics, Institute of Mechanical and Electrical Engineering
University of Southern Denmark
Sønderborg, Denmark

ISBN 978-3-032-32449-8 ISBN 978-3-032-32450-4 (eBook)
https://doi.org/10.1007/978-3-032-32450-4

This Springer imprint is published by the registered company Springer Nature Switzerland AG
The registered company address is: Gewerbestrasse 11, 6330 Cham, Switzerland

This book is dedicated to our families. Here is an opportunity to say special thanks to our families for their concerns and for being the source of inspiration during preparing and writing this book. For sure, producing a comprehensive book is always a challenging experience and this would have not been possible without our family's ongoing understanding and support. Also, the authors would like to thank the groups and individuals for their continuous motivation and support during writing of this book. We would also like to extend our appreciation to the organization and university who supported and contributed to the research in terms of their time and support to formulate our thinking with regard to power system engineering topics.

Preface

The power system structure is a complex framework that comprises multiple elements and subsystems interconnected with each other and operating in synchronization for effective generation, transmission, and distribution within a power system. The system is initiated by power generation through facilities such as power plants, which convert different sources of energy into electrical energy. The generated electrical energy is transmitted from the generating station to the substation via a high-voltage transmission line. At the substation, the level of voltage is stepped down appropriately for further distribution through the low-voltage power network, which will lead to energy delivery for consumers, residents, and industries. Each of the elements in the power system, whether a generator or transformer, plays a vital role in ensuring the reliability and efficient operation of the power system.

This book, *Power System Fundamentals with PowerFactory*, meets that need with remarkable clarity and balance. Written as a complete technical resource, it guides the reader from the basic structure of power systems through single- and three-phase fundamentals, detailed modeling of transformers and synchronous machines, transmission-line characteristics, power-flow and short-circuit calculations, protection principles, and finally to stability and control. What distinguishes this volume is its seamless integration of rigorous theory with hands-on application. Every major topic is reinforced by step-by-step tutorials in PowerFactory. Readers will not only learn *why* a phenomenon occurs but will also gain practical skills to model, simulate, and solve real engineering problems.

Attention is paid to the challenges of energy transition. The text draws extensively on Danish and Nordic grid data, highlighting the rapid growth of wind and solar, the variability of renewable output, and the evolving role of storage and demand response. These examples are especially relevant for engineers working in countries that have already achieved high renewable penetration and are now tackling the next frontier: grid-forming inverters, synthetic inertia, and hybrid AC/DC systems. The inclusion of symmetrical-component analysis, distance and differential protection, governor and excitation control, and equal-area criterion studies equips readers to address both classical and emerging stability issues.

Whether you are a university student building your first power-system model, a practicing engineer responsible for protection coordination or grid planning, or a researcher exploring future power-system architectures, this book provides a clear pathway from theory to practice. Its logical progression, numerous solved examples, PowerFactory screenshots, and end-of-chapter problems make it an ideal textbook as well as a valuable reference for professional development.

Sønderborg, Denmark

Navid Bayati
Mohammed Ali Khan

Competing Interests The authors have no competing interests to declare that are relevant to the content of this manuscript.

Contents

About the Authors

Navid Bayati received his M.Sc. degree in Electrical Engineering from Amirkabir University of Technology, Tehran, in 2017. Since 2019, he has been involved in research with Loughborough University, UK. He graduated with a Ph.D. degree in Electrical Engineering (Power Systems) from the Department of Energy Technology, Aalborg University, Denmark, in November 2020. Then, he was involved as a Postdoctoral researcher at the University of Southern Denmark, since March 2021, on the protection of Microgrids by dSPACE and OPAL-RT. He started his Assistant Professorship in 2022 in power system protection at the University of Southern Denmark. He is currently an Associate Professor at the University of Southern Denmark on power systems. His research interests are power system protection, DC Microgrid, and fault detection and location of renewable energy resource-based systems. He was a recipient of the top 1% reviewer in the world in 2019.

Mohammed Ali Khan is an Assistant Professor at the Centre for Industrial Electronics, University of Southern Denmark. He received his Ph.D. in Power Management of Grid-Connected Distributed Generation from the Advanced Power Electronics Research Laboratory, Department of Electrical Engineering, Jamia Millia Islamia (A Central University), New Delhi, India, in 2021. He was a Postdoctoral Researcher at the Centre for Industrial Electronics, University of Southern Denmark, from February 2023 to January 2025, and at the Department of Electrical Power Engineering, Faculty of Electrical Engineering and Communication, Brno University of Technology, Czech Republic, from December 2021 to January 2023. He also served as a Visiting Researcher at the Center of Reliable Power Electronics, Aalborg University, Denmark, from October to December 2020. He researches focuses on power electronics, intelligent control, and their applications in renewable energy integration, power quality enhancement, multiphysics Modelling, and grid reliability. He has authored several peer-reviewed publications and presented his work at international conferences.

Introduction 1

Abstract

This chapter introduces the fundamental architecture and contemporary challenges of electrical power systems. The chapter further addresses essential supporting functions that underpin reliable operation, including control and protection systems (SCADA and EMS), grid stability and power quality measures (frequency/voltage regulation, load balancing, and demand response), and the integration of variable renewable generation with advanced storage solutions such as battery energy storage systems (BESS) and hydrogen-based technologies. These elements collectively support the transition toward smarter, more resilient, and decarbonized grids. A dedicated statistics section highlights recent global and regional trends.

Keywords

Power system structure • Conventional and renewable generationCables • High-voltage transmission & distribution • Renewable integration and energy storage • Grid statistics and sustainability

1.1 Structure of Power System

The power system structure is a complex framework that comprises multiple elements and subsystems interconnected with each other and operating in synchronization for effective generation, transmission, and distribution within a power system, as represented in Fig. 1.1.

The system is initiated by power generation through facilities such as power plants, which convert different sources of energy into electrical energy. The generated electrical energy is transmitted from the generating station to the substation via a high-voltage transmission line. At the substation, the level of voltage is stepped down appropriately for further distribution through the low-voltage

N. Bayati and M. A. Khan, *Power System Fundamentals with PowerFactory*,
https://doi.org/10.1007/978-3-032-32450-4_1

Fig. 1.1 General structure of power systems

power network, which will lead to energy delivery for consumers, residents, and industries. Each of the elements in the power system, whether a generator or transformer, plays a vital role in ensuring the reliability and efficient operation of the power system. The power system can be generally divided into three main stages.

1.1.1 Generation System

The generation system is the primary element of the electrical power network, as it is responsible for converting various energy sources into electrical power. The sources can be either conventional or renewable power sources. Over recent years, generation units have evolved by incorporating more renewables into the power system to reduce the carbon footprint and improve sustainability. The advancement in energy storage technologies and hydrogen-based power solutions, along with smart grids, have been at the forefront of improving grid stability and reliability even when the energy demand has been growing exponentially. Further to explain the power sources in detail, the generation system can be classified as.

1.1.1.1 Conventional Source-Based Power System

Thermal power plants: Thermal power plants convert the thermodynamic energy, which is generated by burning fossil fuel, into mechanical energy and further into electrical energy. The process is initiated by bringing fuel for heating up the boiler and producing steam, which is stored under high pressure and can be used to drive a turbine. Pressurized steam is used for rotating the turbine blades, converting thermal energy into mechanical energy, whereas the rotating turbine is exposed to a dynamo for generating electrical energy. Thermal power plants have been a primary source of energy generation for a long time; however, they have been contributing significantly to environmental challenges such as global warming and greenhouse gases. In the past, there has been an effort made to reduce emissions through carbon capture and heat recovery.

Nuclear Power Plants: Nuclear Power Plants are specialized facilities that generate electricity through controlled nuclear reactions. They primarily utilize nuclear fission, a process in which atoms of heavy elements like uranium or plutonium are split into smaller nuclei, releasing substantial amounts of heat. This thermal energy is used to produce steam, which powers turbines connected to generators.

The core of this operation takes place inside a nuclear reactor, designed to sustain and regulate the fission process. To ensure safety, these plants are equipped with robust cooling systems and containment structures that manage radioactive materials and mitigate the risk of accidents.

Nuclear energy is considered a low-carbon source of power, offering potential benefits for reducing greenhouse gas emissions. However, it also presents challenges related to radioactive waste disposal, safety protocols, and the risk of catastrophic incidents.

Hydroelectric Power Plants: Hydroelectric power plants generate electricity by harnessing the energy of flowing water. Typically, a dam is built across a river to create a reservoir, which stores water at a higher elevation. When the water is released, it flows through turbines, causing them to spin. These turbines are connected to generators that convert mechanical energy into electrical energy. Hydroelectric power is considered a renewable energy source because it relies on the natural water cycle, and it can produce large amounts of power with low greenhouse gas emissions. However, the construction of dams can have environmental and social impacts, such as altering ecosystems and displacing communities.

1.1.1.2 Renewable Source-Based Power System

Solar Photovoltaic (PV) Systems: Photovoltaic (PV) technology harnesses sunlight to generate electricity through the photovoltaic effect, where solar energy is directly converted into electrical energy using semiconductor materials. PV systems, commonly seen in solar panels, are a key component of renewable energy solutions due to their scalability, low environmental impact, and declining cost. Widely deployed in residential, commercial, and utility-scale applications, PV technology contributes significantly to reducing dependence on fossil fuels and mitigating greenhouse gas emissions. Continuous advancements in materials, such as silicon and perovskites, and improvements in efficiency and storage integration are enhancing the performance and adoption of PV systems worldwide.

Wind Turbines: Wind is a renewable and sustainable source of energy that has been harnessed for centuries, and in modern times, it plays a critical role in the global transition toward cleaner power systems. Wind energy is generated by converting the kinetic energy of moving air into mechanical power using wind turbines, which is then transformed into electricity. The development of wind technology has advanced rapidly, with improvements in turbine efficiency, size, and grid integration capabilities. As a result, wind power has become one of the fastest-growing energy sources worldwide, contributing significantly to reducing greenhouse gas emissions and dependence on fossil fuels. Its scalability, from small distributed systems to large offshore wind farms, makes it a versatile solution for meeting both local and national energy demands.

Geothermal Power: Geothermal energy is a renewable and sustainable energy source derived from the natural heat of the Earth's interior. This thermal energy originates from the decay of radioactive elements and the residual heat from the planet's formation, which accumulates in rocks and fluids beneath the Earth's crust. By tapping into this heat through wells and specialized technology, geothermal systems can provide a constant and reliable supply of electricity and direct heating. Unlike solar or wind energy, geothermal is largely unaffected by weather or seasonal changes, making it a stable contributor to a diversified energy portfolio. As global demand for clean and dependable energy grows, geothermal energy presents a promising solution for reducing greenhouse gas emissions and supporting energy security.

1.1.2 Transmission

After generation, electricity is transported over long distances via high-voltage transmission lines. High Voltage Direct Current (HVDC) and High Voltage Alternating Current (HVAC) are two fundamental technologies used for the transmission of electrical power over long distances. HVAC, the traditional method, is widely used for local and regional power distribution due to its ease of voltage transformation and established infrastructure. However, over very long distances or in underwater and underground applications, HVDC becomes more efficient due to lower line losses and better controllability. HVDC systems enable the interconnection of asynchronous power grids and facilitate the integration of large-scale renewable energy sources. Both technologies play complementary roles in modern power systems, supporting reliable, efficient, and flexible electricity transmission networks. Key components of this stage include:

Step-up Transformers: Increase voltage levels for efficient transmission. A step-up transformer is an essential electrical device used to increase voltage levels from a lower value at the primary winding to a higher value at the secondary winding, while proportionally decreasing the current. This type of transformer plays a critical role in power transmission systems, especially in the generation and delivery of electricity over long distances. By stepping up the voltage, electrical power can be transmitted more efficiently with reduced energy losses due to lower current flow in the transmission lines. Step-up transformers are commonly found in power plants, where they boost the generator output voltage to high transmission levels, facilitating the economical and reliable distribution of electricity across vast networks.

Transmission Lines: They are vital components of power systems, enabling the efficient transfer of electrical energy over long distances. In High Voltage Alternating Current (HVAC) systems, transmission lines operate using sinusoidal AC voltages, which are widely used due to ease of voltage transformation and well-established infrastructure. However, HVAC systems are subject to reactive power losses and line charging effects over long distances. In contrast, High Voltage Direct Current (HVDC) transmission lines use direct current to transfer electricity, offering advantages such as lower line losses, better controllability, and the ability to connect asynchronous power grids. HVDC is particularly suitable for long-distance and submarine transmission, as well as for integrating renewable energy sources. Together, HVAC and HVDC technologies form the backbone of modern power transmission networks, each optimized for specific technical and economic conditions.

Substations: A substation is a critical component of the electrical power system, serving as an interface between different voltage levels and playing a vital role in the transmission and distribution of electricity. It functions as a node where electrical power is transformed from high to low voltage or vice versa, ensuring efficient and safe delivery from generation sources to end-users. Substations are equipped with various electrical devices such as transformers, circuit breakers, isolators, and protective relays, which help control, protect, and regulate the flow of electrical energy. Depending on their function and voltage levels, substations can be categorized as transmission, distribution, or switching substations. Their strategic placement and design are essential for maintaining system reliability, voltage stability, and power quality within the grid.

Grid Interconnections: Grid interconnection refers to the integration of multiple power systems or energy sources into a unified electric grid, enabling efficient, reliable, and flexible electricity transmission and distribution. As the global energy landscape shifts toward more decentralized and renewable energy sources—such as solar, wind, and hydrogen-based systems—grid interconnection becomes essential for balancing supply and demand, enhancing energy security, and improving system resilience. This interconnected framework allows for power sharing across regions, supports the integration of variable renewable generation, and facilitates the development of smarter, more adaptive grids.

1.1.3 Distribution

The transmission voltage is stepped down and distributed to end users. Medium and low-voltage lines, both overhead and underground, form the backbone of the electrical distribution network, delivering electricity from substations to end users such as homes, businesses, and industries. These networks typically operate at voltages ranging from 1 to 36 kV for medium voltage and below 1 kV for low voltage. Overhead lines are widely used due to their lower installation cost and ease of maintenance, while underground cables are preferred in urban areas for aesthetic, safety, and space-saving reasons. Together, these systems ensure reliable and efficient energy delivery, playing a crucial role in modern power distribution infrastructure. Components of the distribution system include:

Step-down Transformers: A step-down transformer is an electrical device used to reduce the voltage from a higher level on the primary side to a lower level on the secondary side while maintaining the same frequency. It operates on the principle of electromagnetic induction and is commonly employed in power distribution systems to safely deliver electricity to residential, commercial, and industrial users. By lowering high transmission voltages to usable levels, step-down transformers play a crucial role in improving energy efficiency, ensuring equipment compatibility, and enhancing safety. Their design typically includes a greater number of windings on the primary coil than on the secondary, which facilitates the voltage reduction process.

Distribution Lines: Distribution lines are a crucial part of the electrical power system responsible for delivering electricity from substations to end consumers, including residential, commercial, and industrial users. These lines operate at lower voltages compared to transmission lines and are designed to efficiently distribute power over relatively short distances while maintaining reliability and safety. Typically spanning overhead or underground, distribution lines form the final stage of the electrical supply chain, ensuring that electricity reaches homes and businesses in a usable form. Their design and maintenance are vital for minimizing losses, preventing outages, and supporting the integration of distributed energy resources such as solar panels and electric vehicles.

Distribution Substations: Distribution substations play a crucial role in the electrical power system by serving as the key interface between the high-voltage transmission network and the lower-voltage distribution network that delivers electricity to end consumers. These substations step down the voltage from transmission levels to distribution levels through transformers, enabling safe and efficient delivery of electrical power to residential, commercial, and industrial users. Besides voltage transformation, distribution substations also help in controlling power flow, protecting the distribution network through circuit breakers and protective relays, and improving system reliability by isolating faults. Their strategic placement and design are essential for maintaining a stable and continuous power supply within local distribution areas.

1.1.4 Other Key Aspects of Power System Structure

Control & Protection Systems: These systems are critical components in modern power networks, ensuring reliable, efficient, and safe operation. Supervisory Control and Data Acquisition (SCADA) systems provide real-time monitoring and control of electrical grid elements, enabling operators to remotely manage and respond to grid conditions. Energy Management Systems (EMS) optimize the generation, distribution, and consumption of power by analyzing data and coordinating system operations to enhance grid stability and efficiency. Protection devices such as relays, circuit breakers, and fuses serve as the first line of defense against faults and abnormal conditions by quickly detecting and isolating faulty sections, thus preventing equipment damage and maintaining system integrity. Together, these systems form an integrated framework that supports resilient and automated power system operation.

Grid Stability & Power Quality: Stability and power quality are critical aspects of modern electrical systems, ensuring reliable and efficient delivery of electricity. Frequency and voltage regulation maintain system stability by keeping the grid's frequency and voltage within prescribed limits, preventing equipment damage and power interruptions. Load balancing involves distributing electrical demand evenly across the network to avoid overloads and voltage fluctuations, while demand response programs actively adjust consumer usage patterns in response to grid conditions, helping to smooth peak loads and enhance overall system resilience. Together, these mechanisms support a stable, high-quality power supply essential for both consumers and grid operators.

Integration of Renewable Energy & Storage: The integration of renewable energy with advanced storage solutions is transforming modern power systems, enabling greater flexibility, reliability, and sustainability. Smart grids and microgrids play a crucial role in intelligently managing distributed energy resources, balancing supply and demand locally, and enhancing resilience against outages. Battery Energy Storage Systems (BESS) provide fast, efficient, and scalable storage, allowing excess renewable generation to be stored and dispatched when needed, thus smoothing variability and supporting grid stability. Complementing these, hydrogen-based energy storage offers long-duration, large-scale storage by converting surplus renewable electricity into hydrogen through electrolysis, which can be stored and later used for power generation, transportation, or industrial applications, making it a promising solution for deep decarbonization and energy sector coupling. Together, these technologies form a synergistic framework essential for a reliable and clean energy transition.

1.1.5 Power System's Statistics

The power generation strategies of the majority of countries have been changed over the last few years to a cost-effective and green energy policy. This green transition with ambitious renewable energy goals and enhanced energy efficiency provides a high level of security of energy supply, while reducing pollution and fossil fuel dependency. Worldwide, the renewable power capacity additions set a record in 2023 with 473 GW of new installed capacity, with a 54% increase compared to 2022. Therefore, the total global renewables capacity in 2023 increased by 14%, from 3391 GW in 2022 to 3865 GW in 2023. In this case, China represented the largest market for solar PV by 63%, onshore wind by 66%, offshore wind by 65%, and hydropower by 44% in 2023, due to the decline in the global weighted average costs for these technologies. The total renewable power generation in the world from 2000 to 2023 had an estimated saving of around USD 409 billion in fuel costs in the power sector. On the other hand, due to the fluctuation of renewable energy production, the battery storage implementation has a wide interest with annual capacity increasing from 0.1 GWh in 2010 to 95.9 GWh in 2023, and the costs of battery storage projects declined by 89% during these years [1]. In summary, the technologies with the highest share of new capacity installations by region in 2023 were PV and onshore wind. The PV system has the largest share of deployment for all regions except Africa, and onshore wind was the second most deployed technology. Africa had a more diverse energy implementation with the same share for onshore wind and hydropower, at 34%. The detailed net non-renewable and renewable generation capacities are presented in Tables 1.1 and 1.2, respectively [2].

For example, in Denmark, greenhouse gas emissions have been reduced by more than 30% since 1990, and 40% in 2020. Denmark has a high share of non-hydro renewable energy generation in the world, with electricity production for 81.4% of the Danish domestic electricity supply in 2022 as opposed to 71.9% in 2021. Wind power contributed 53.6%, biomass 18.9%, and solar energy and biogas the remaining 8.8% [3]. Therefore, understanding the current and future energy policies of Denmark can help in the understanding of the future power system structure [4]. In Fig. 1.2, the daily CO_2 emissions during the production of electricity in the Danish power grid are illustrated [5]. As can be seen, the CO_2 emissions change from around 30 kg/GWh on some days to around 300 kg/

Table 1.1 Net non-renewable generation capacity

Country	Non-renewable net generation (MW)	Country	Non-renewable net generation (MW)
AL	97	IE	7,675
AT	8,637	IT	61,635
BA	2,323	LT	2,525
BE	14,093	LU	91
BG	8,741	LV	1,157
CH	7,789	ME	210
CZ	15,292	MK	1,240
DE	89,921	NL	23,630
DK	5,934	NO	1,645
EE	1,555	PL	34,139
ES	44,994	PT	8,320
FI	8,783	RO	5,961
FR	85,769	RS	7,328
GE	1,189	SE	15,900
GR	9,379	SI	2,776
HR	1,315	UA	15,798
HU	6,392		

Table 1.2 Net renewable generation capacity

Country	Renewable net generation (MW)	Country	Renewable net generation (MW)
AL	2,208	IE	2,152
AT	15,989	IT	34,061
BA	1,843	LT	1,619
BE	12,706	LU	532
BG	6,008	LV	2,014
CH	18,133	ME	767
CZ	4,354	MK	710
DE	142,839	NL	32,452
DK	11,233	NO	38,633
EE	987	PL	21,076
ES	65,762	PT	11,823
FI	10,531	RO	10,396
FR	58,044	RS	3,075
GE	3,425	SE	31,000
GR	12,618	SI	1,848
HR	3,202	UA	4,637
HU	3,971		

GWh at another time. In Fig. 1.3, the non-renewable and solar generation of the Danish power grid during 2024 for all days has been shown. As could be seen, the maximum power production by solar generation belonged to the beginning of April to the end of August, and at the same time, the power production by non-renewable generations, based on fuels, decreased. Therefore, increasing the renewable energy generation will decrease the power production by fuel-based generation units, as always, renewable energy generation has a higher priority.

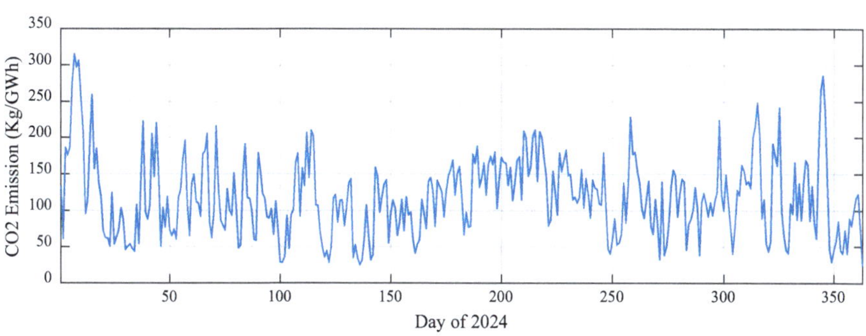

Fig. 1.2 CO_2 emission of the Danish power system during 2024

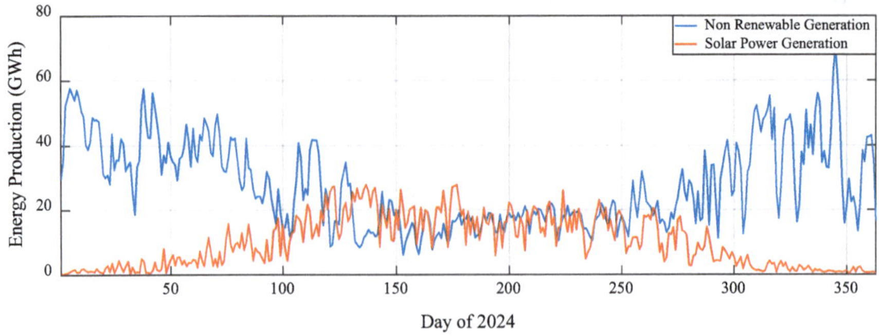

Fig. 1.3 Non-renewable and solar generation of the Danish power grid during 2024

The 2022 Nordic and Baltic Grid Disturbance Statistics by Regional Group Nordic [6] gives an overview of the grid disturbances and energy not supplied (ENS) in the Nordic and Baltic regions for the transmission grids of 100–420 kV. In 2022, disturbances had a much lower impact on the end-users compared to the average 10-year period of 2013–2022. ENS was 54% in the Baltic and 55% in Nordic countries of the corresponding 10-year annual averages. In 2022, the number of faults and the number of disturbances were about 5% smaller in Baltic and about 10% higher in Nordic, which caused ENS to be only half in Baltic and at the same level in Nordic. The largest share of ENS in 2022 was for overhead line with 99% in Finland, 59% in Iceland, 57% in Sweden, 56% in Lithuania, and 45% in Norway. The statistics for the number of disturbances and ENS for Nordic and Baltic countries are shown in Table 1.3. As can be seen, Iceland has the lowest number of disturbances; however, at the same time, it has the highest rate of ENS among all Nordic and Baltic countries.

Table 1.3 The number of disturbances and ENS in 2022 and the corresponding annual averages for the 10-year period 2013–2022

Country	Number of disturbances		ENS (MWh)	
	2022	Average 2013–2022	2022	Average 2013–2022
Estonia	144	165	50	74
Latvia	118	126	13	78
Lithuania	162	151	39	37
Denmark	80	65	37	34
Finland	572	418	125	256
Iceland	58	44	1269	797
Norway	292	301	418	2151
Sweden	426	438	650	1343

1.2 Outline of the Book

This book is composed of 7 chapters and describes the fundamental, main components, operation, protection, and control of power systems. Moreover, this book covers the theory of different topics, and the practical approach by using PowerFactory—DIgSILENT to provide the readers a deep appreciation of power system analysis.

Chapter 2 discusses the fundamentals of power system analysis from single to three-phase power systems. It provides the basic understanding for readers about the AC voltage and current, active and reactive power, and per-unit calculations. In this chapter, the basics of PowerFactory—DIgSILENT will be covered to provide a getting-started tutorial, in addition to providing simple modeling and analysis examples.

Chapters 3 and 4 provide the background on working principles and modelling of main components in the power grid. In Chap. 3, the modelling and theory behind the transformers and synchronous machines will be covered from ideal to accurate modelling of them. In Chap. 4, the overhead transmission lines and cables will be introduced. The modelling, impacts, and challenges of these components will be illustrated. At the end of each chapter, the practical PowerFactory—DIgSILENT modelling of all components, with a deep tutorial on them, will be covered.

Chapter 5 will cover the theory behind the power flow and short-circuit calculations for power systems with different methods. The understanding of this chapter will help to discuss the protection and control units in Chaps. 6 and 7. In Chap. 6, different protection devices will be introduced, and their settings will be covered, and practically will be implemented at the end of Chap. 6 with the help of PowerFactory – DIgSILENT. Chapter 7 also addresses the control and stability of power systems with different control units.

References

1. IRENA (2024) Renewable power generation costs in 2023. International Renewable Energy Agency, Abu Dhabi
2. Entsoe, Statistical Factsheet (2023) Provisional values as of April 2024
3. Danish Energy Agency (2022) Data, tables, statistics, and maps energy statistics
4. Gharehpetian GB, Zolfaghari M, Bayati N (2024) Future power system elements, challenges, and solutions. Elsevier
5. Energi Data Service. Power system right now. www.energidataservice.dk
6. ENTSO-E (2023) 2022 Nordic and Baltic grid disturbance statistics

Fundamentals of Power Systems

2

Abstract

This chapter has established the foundational principles of power system analysis by examining both single-phase and three-phase AC networks. In single-phase systems, the sinusoidal variation of voltage and current, RMS quantities, instantaneous power, and the decomposition into active and reactive power components were discussed in detail, along with the importance of power factor in system efficiency. The advantages of three-phase systems for bulk power transfer were then presented, including star and delta configurations, line-to-phase relationships, balanced power calculations, per-unit system normalization for simplifying multi-component network studies, and symmetrical components for handling unbalanced conditions and fault analysis. Finally, a practical introduction to PowerFactory—DIgSILENT was provided, demonstrating step-by-step modeling of a simple two-bus system, data entry for components such as transformers and loads, execution of balanced AC load-flow calculations, and visualization of results. These concepts and tools form the essential basis for subsequent chapters on power system components, load flow, short-circuit analysis, protection, and stability studies.

Keywords

Three phase systems • Power factor • Delta and star connection

This chapter presents the fundamental concepts of power systems, including generation, transmission, and distribution. The chapters include both three-phase and single-phase power systems, while underlining their characteristics, organization, and application. For the single-phase system, the behavior of voltage, current, along with power factor and load characteristics (resistive, inductive, and capacitive) is presented. Whereas for the three-phase system, discussion is carried out

N. Bayati and M. A. Khan, *Power System Fundamentals with PowerFactory*,
https://doi.org/10.1007/978-3-032-32450-4_2

regarding star and delta connections along with line-to-phase relationships. The chapter aims to present a clear outline of power flow from generation to distribution end, with an aim to establish a foundation for power flow, system stability, and renewable energy integration.

2.1 Introduction

Economic growth in the modern world is highly attributed to the development of power system infrastructures around the world. As recently as 2023, the global installed capacity was reported to be about 9.4 terawatts (TW), with a prediction of 15.9 TW by 2030 []. With the extensive power system in operation, the system needs to be organized in a three-phase as well as a single-phase AC network. A single-phase system is usually used for a low-power load, which can be either household or commercial. Such loads are either 110 V or 240 V, based on the region, and are used for lighting, heating, or the operation of small appliances. However, single-phase systems are not ideal for powering heavy loads, and their efficiency also drops significantly if they need to be transmitted over a long distance. For such cases, three-phase systems are used, and they are common for industrial loads as well as transmission over long distances. The three phases operate at a phase difference of 120°, which accounts for consistent power delivery with better efficiency. This results in transferring more power with lower losses.

Global trends indicate a rapid increase in installed power capacity. It was reported that alone in 2023, there was an increase of 13.9% of power capacity. Out of the 473 GW installed, 246 GW was generated through solar, whereas 116 GW was generated through wind energy. Whereas to maintain the goal for achieving net neutrality, the estimated global installation needs to be increased to 11 TW by 2030. Furthermore, to understand the nature of these installations, it is essential to have a good insight regarding the fundamentals, from single-phase to three-phase systems, as they are crucial for the efficient and reliable operation of power networks.

This chapter aims to discuss the fundamentals of the power system, along with the essential knowledge regarding single and three-phase power systems, along with the application of each system and its best use, as well as tradeoffs. Core electrical concepts from single-phase to three-phase networks can have multiple concerns, ranging from line/phase relationships, Y/Δ connections, balanced vs. unbalanced operation, as well as power measurement issues. The overall outlook for the EU market is summarized in Table 2.1.

Table 2.1 Overall outlook of the EU market

Country	Generated energy	Renewable share	Major green sources (approx.)
Germany	Around 431.7 TWh (2024)	Around 59.0% is from renewables (2024)	Onshore wind about 111.9 TWh Solar about 63 TWh
Spain	Around 135 TWh (2023)	Around 50.3–50.4% is from renewables (2023)	Wind: 23–24%, Solar PV: 14%, Hydro: 9–10% are major;
Denmark	NA	Among the highest in the EU for the share of renewables (wind), approx 80–90% in some months	Very strong wind power sector (onshore/offshore)
France	About 495 TWh (2023)	About 27–30% of renewables (2023)	Renewable share of about 27%
Italy	About 312,000 GWh (2024)	About 41.2% of electricity (2023)	Renewables include hydro, solar PV, and wind

2.2 Single-Phase Power Systems

The residential and small commercial applications commonly use a single-phase power system, which comprises an AC electrical distribution system. A single-phase system comprises a live wire (phase line) and a neutral wire. The phase wire supplies current to the load, whereas the neutral wire provides a return path and completes the circuit. The power delivery in this configuration is through a single AC waveform with a frequency ranging from 50 to 60 Hz, depending upon the region of application.

In a single phase, the flow of current is alternating in nature, which makes the transmission of energy efficient over long distances; however, it also means that the power delivery is not constant for the load throughout. The instances when the instantaneous value is zero can affect the operation of various loads significantly.

Single-phase systems are primarily used in places where the power requirement is limited, such as residential loads, small businesses, and commercial establishments. The power requirement is within the range, making it efficient and cost-effective.

Even though single-phase systems are appropriate for low-power applications, they have constraints after a certain power limit. The major drawback is the lack of a rotational magnetic field, which is crucial for the effective operation of electric motors. Subsequently, motors when operating in single-phase systems frequently require supplementary components, such as capacitors or auxiliary windings, for starting and operating them effectively. Furthermore, inconsistent power delivery can lead to problems in systems that insist on steady power. Further, such fluctuations can cause vibrations, noise, and potential wear on mechanical components.

Hence, single-phase power systems form the basis of electrical distribution in residential and light commercial sectors. Their cost-effectiveness and suitability for low-power applications make them an ideal choice for everyday use. Even though they may not be the best alternative for high-power applications.

2.2.1 AC Voltage and Current

For a single-phase AC system, the voltage and current vary with time in a sinusoidal form, which can also equate to a change in the direction of the magnitude at different instances. AC is also much more efficient for long-distance transmission of electrical energy, unlike DC, which happens to have constant current and voltage. AC can also be transformed into other voltage levels through transformers, which are a fundamental part of distribution networks [1].

The instantaneous voltage $v(t)$ in a single-phase AC system can be represented as:

$$v(t) = V_m \sin(\omega t + \phi_v) \tag{2.1}$$

where, the peak voltage is denoted by V_m, ω represents the angular frequency ($\omega = 2\pi f$, f in Hz), t denotes time, and ϕ_v is the phase angle for the voltage. Correspondingly, the instantaneous current is denoted by $i(t)$, and can be expressed as:

$$i(t) = I_m \sin(\omega t + \phi_i) \tag{2.2}$$

where, the current peak is represented by I_m and ϕ_i is the current phase angle. The phase difference $\phi = \phi_v - \phi_i$ between the phases of voltage and current establishes whether the load is resistive, inductive, or capacitive in nature. For instance, in the case of purely resistive loads, the phase difference is zero ($\phi = 0$), whereas in the case of purely inductive loads, the current will lag voltage by 90° ($\phi = 90°$), and for a purely capacitive load, current leads voltage by 90° ($\phi = -90°$) [2].

The root mean square (RMS) values of voltage and current are used for analysis, as they represent the equivalent DC values that would deliver the same power for the resistive load:

$$V_{rms} = \frac{V_m}{\sqrt{2}}, I_{rms} = \frac{I_m}{\sqrt{2}} \tag{2.3}$$

Further, instantaneous power $p(t)$ delivered to the load is denoted as:

$$p(t) = v(t) \cdot i(t) \tag{2.4}$$

In case of the resistive loads, the power is always positive, whereas for non-linear loads, the instantaneous power fluctuates around the axis, which gives rise to reactive power. Reactive power does not contribute to net energy transfer but certainly affects the design and operation of the system [3]. Comprehension of the relations between AC voltage, current, and phase angle is crucial for investigating circuits, designing electrical equipment, and ensuring efficient energy usage.

2.2.2 Active Power

For a single-phase AC system, active power is correspondent as real power, as it is the power that is translated from the electrical energy in other forms, such as heat, light, and mechanical energy. Such energy can be measured in watts, represented as W. In AC systems, voltage and current operate in a sinusoidal nature with time and can be expressed as:

$$\begin{aligned} v(t) &= V_m \sin(\omega t) \\ i(t) &= I_m \sin(\omega t - \phi) \end{aligned} \tag{2.5}$$

where, V_m and I_m denotes the peak values of voltage and current, respectively, whereas $\omega(2\pi f)$ represents the angular frequency, f is representing the frequency of the system, and ϕ is the phase angle between the current and voltage. The instantaneous power consumed by the load is a product of instantaneous voltage and current, which can be represented as:

$$p(t) = v(t) \times i(t) \tag{2.6}$$

Further on, substituting the extended expression as:

$$p(t) = \frac{V_m I_m}{2}[\cos(\phi) - \cos(2\omega t - \phi)] \tag{2.7}$$

It can be observed that the equation comprises two components, the term $\frac{V_m I_m}{2}\cos\phi$ denotes active power, whereas the second term $\frac{V_m I_m}{2}\cos(2\omega t - \phi)$ represents reactive power.

The active power can be expressed in RMS as:

$$P = V_{rms} I_{rms} \cos\phi \tag{2.8}$$

Here, V_{rms} and I_{rms} represent the RMS values of voltage and current, and $cos\phi$ represent the power factor, indicating the efficiency of power used. Power factor 1 represents that all the supplied power is active and is being successfully converted into work. It is like operating in purely resistive circuits where voltage and current are in phase. Nonetheless, for inductive or capacitive loads, the current and voltage operate out of phase, resulting in a lower power factor with a reduced active power for the constant apparent power level.

According to **Elements of Power System Analysis** and **Power System Analysis and Design**, comprehending active power is critical for designing and analyzing power systems as it correlates to the performance and efficiency of electrical networks. Further discussions on standards can be found in **IEEE Std 1459-2010**, which emphasizes that active power expands as the foundation of modern power engineering.

2.2.3 Reactive Power

For a single-phase AC system, the oscillating power translates into reactive power, particularly for inductive and capacitive loads. This power is associated with activities such as energy storage in electrical and magnetic fields and doesn't translate into other forms of energy, such as light, heat, or mechanical action.

For a resistive load, when a voltage is applied, the current remains in phase with the voltage; hence, all the power is transferred toward the work done, however in the case of capacitive and inductive loads, there is a certain leading or lagging phase shift, respectively, between the voltage and the current, with a phase shift of ϕ. This phase shift is responsible for the reactive power.

As mentioned in the section before $\frac{V_m I_m}{2}\cos(2\omega t - \phi)$ represents reactive power in the overall power equation. This also establishes that power oscillates due to the cosine term, and only part of it ($\cos\phi$) represents average, or real, power. The reactive power for the RMS term can be represented as:

$$Q = V_{rms} I_{rms} \sin\phi \tag{2.9}$$

Reactive power is measured in volt-ampere reactive (VAR). Further, the apparent power (S) accounts for the total power supplied, is the vector sum of real and reactive power:

$$\begin{aligned} S &= V_{rms} I_{rms} \\ S^2 &= P^2 + Q^2 \end{aligned} \tag{2.10}$$

These relationships frequently use the power triangle, where P is the horizontal component, Q is the vertical component, and S is the hypotenuse. The PF of the system is given by:

$$PF = \cos\phi = \frac{P}{S} \tag{2.11}$$

A low power factor implies a high reactive power, which leads to increased current flow and causes higher losses. Further, power factor improvement, in general, using capacitors or synchronous condensers, is utilized to reduce reactive power demand and improve productivity.

2.3 Three-Phase Power System

The three-phase power system is primarily used for the generation, transmission, and distribution of power around the globe. It is more stable and efficient when compared to the single-phase power system. A three-phase power system comprises three AC voltage signals of identical amplitude and frequency, just operating at a phase shift of 120º from each other. By operating the power system as such, the power flow is way more balanced with reduced pulsing as well as better performance for industrial loads as well as grid.

The most substantial benefit of the three-phase system is its ability to deliver constant power. In case of single-phase systems, the instantaneous power fluctuates twice per cycle, whereas during a balanced three-phase system, the sum of the instantaneous powers for the three phases remains constant over time. The continuous power delivery affects the efficient operation of industrial loads, improves transmission efficiency, and reduces mechanical vibration. Furthermore, three-phase systems are more economical as they expect fewer conductor material for the same transmitted capacity when compared with single-phase systems. This effectiveness is an essential factor for large-scale operation, as the material costs and energy losses have significant economic consequences.

Three-phase power remains essential for the function of large industrial plants and high-voltage transmission networks, as it forms the modern renewable energy systems. For instance, wind turbine generators and large-scale photovoltaic inverters are essentially three-phase devices planned to supply directly into three-phase grids. It safeguards reliable power delivery and assists in the integration of renewables with the existing grid.

The phase sequence indicates the order in which the phases reach their peak values. A standard sequence is indicated as ABC, which denotes that phase A leads B by 120°, and leads C by 240°. Reversing any of the two-line connections leads to a negative sequence (ACB), which impacts the direction of rotation in three-phase motors. The correct phase sequence is therefore essential in industrial applications and protection coordination. Further, the system is considered balanced when all phases have equal magnitudes, and phase displacement between each pair of phases is exactly 120°. Any deviation from these ideal conditions signifies an unbalanced system, which can lead to increased copper and iron losses, overheating of machines, torque pulsations, and mechanical vibration.

A thorough understanding of balanced and unbalanced operation is critical for load-flow analysis, fault calculations, stability assessments, and harmonic studies.

In brief, a strong understanding of three-phase operational principles, including phase relationships, system configurations, and power flow characteristics, forms the substance for advanced studies in grid stability and smart grid formation and operation.

2.3.1 Three-Phase Voltage and Current

A balanced three-phase system consists of three sinusoidal voltages of equal magnitude and frequency, phase-shifted by 120°:

Each of the phrases can be represented as:

$$\begin{aligned} v_a &= V_m \sin(\omega t) \\ v_b &= V_m \sin\left(\omega t - 120^\circ\right) \\ v_c &= V_m \sin\left(\omega t - 240^\circ\right) \end{aligned} \tag{2.12}$$

where, the voltage peak is represented by V_m, and $\omega = 2\pi f$ is the angular frequency with a system frequency f. Such systems are arranged in one of two main connection types, star (Y) or delta (Δ). During star connection, the three windings are connected at a common neutral point, which is grounded, and the association between the line voltage (V_L) and the phase voltage (V_P) is:

$$V_L = V_P/\sqrt{3} \tag{2.13}$$

This arrangement enables the utilization of the neutral wire for providing both line-to-line and line-to-neutral voltage. Whereas in a delta connection, each end is connected to the other, which leads to a correlation between the line current (I_L) and phase current (I_P) as:

$$I_L = I_P/\sqrt{3} \tag{2.14}$$

whereas the $V_L = V_P$. Delta arrangements are specifically used in industrial applications where balanced loads and higher starting torques are required, particularly in the case of an induction motor.

2.3.2 Three-Phase Active and Reactive Power

It has already been well established that three-phase power systems offer substantial advantages over single-phase systems. When the voltages are equal in magnitude and phase shift of 120°, the total instantaneous power provides a minimum of torque pulsations in industrial loads, further improving overall system performance.

The power equations for a balanced three-phase system can be expressed as follows:

$$\begin{aligned} P &= \sqrt{3}\, V_L\, I_L \cos\phi \\ Q &= \sqrt{3}\, V_L\, I_L \sin\phi \\ S &= \sqrt{3}\, V_L\, I_L \end{aligned} \tag{2.15}$$

These relationships are fundamental for estimating power transfer, system losses, and power factor analysis for a three-phase network.

2.3.3 Per-Unit Calculations

The per unit (p.u.) system aims at reducing the complexity of the power system by normalizing the unit and eliminating transformation ratios and other considerations across the networks. For the implementation of such methods, the electrical quantities are stated in fractions with a particular base value, S_{base} is the value for the base power, whereas V_{base} is the base value for the line-to-line voltage. The base current and impedance can be calculated as:

$$I_{\text{base}} = \frac{S_{\text{base}}}{\sqrt{3}\, V_{\text{base}}} \text{ and } Z_{\text{base}} = \frac{V_{\text{base}}^2}{S_{\text{base}}} \tag{2.16}$$

The basic principle of the per-unit system is to consider any actual quantity X_{actual} and factor it with a base value as follows:

$$X_{\text{p.u.}} = \frac{X_{\text{actual}}}{X_{\text{base}}} \tag{2.17}$$

By doing so, the parameters such as voltages, currents, impedances, and powers become dimensionless, and they are all in a numerical range (usually 0.1–1.2 p.u.). Another advantage that can be noted is that per unitization removes the issues with the transformation ratio and simplifies the analysis process for the circuit with transformers. This removes the impedance scaling and reduces the computational errors. For the elements that are expressed with different base values, the per-unit impedance can be transferred using the change-of-base formula

$$Z_{\text{p.u., new}} = Z_{\text{p.u., old}} \left(\frac{V_{\text{base, old}}}{V_{\text{base, new}}} \right)^2 \left(\frac{S_{\text{base, new}}}{S_{\text{base, old}}} \right) \tag{2.18}$$

This allows equipment with distinct ratings to be merged seamlessly into a system model. For three-phase systems, the per-unit system uses three-phase power bases such as:

$$S_{3\phi,\,\text{base}} = \text{MVA base},\ V_{\text{LL, base}} = V_{\text{rated}}, \tag{2.19}$$

and similarly, the base current becomes

$$I_{\text{base}} = \frac{S_{\text{base}}}{\sqrt{3}V_{\text{LL, base}}} \tag{2.20}$$

As the per-unit eliminates large numerical dissimilarities amongst components like lines, transformers, generators, inverters, and loads, they tend to improve numerical stability for the software-based analysis, particularly in the case of load-flow analysis, short-circuit and transient stability studies, as well as EMT simulations. Furthermore, regular per-unit reactance values for a synchronous generator (e.g., $X_d \approx 1.0\,\text{p.u.}$, $X_d' \approx 0.3\,\text{p.u.}$, $X_d'' \approx 0.15\,\text{p.u.}$) represent the usefulness of p.u. system as it makes machine constraints directly comparable across different power ratings. In the case of power electronic converters, per-unit is essential as it normalizes control parameters, confirming stable numerical behavior during switching events, and ensuring integrated modeling across altered voltage levels. Because of this, the per-unit system remains necessary in all future power system studies, from steady-state to dynamic and transient analysis in modern power systems.

2.3.4 Three-Phase Components

Three-phase components are most effectively analyzed using the concept of symmetrical components, which consist of any unbalanced set of three-phase phasors as well as three balanced sequence sets, positive, negative, and zero sequence, using the transformation. If the original phase quantities are V_a, V_b, and V_c, the symmetrical components can be obtained through the following:

$$\begin{bmatrix} V_0 \\ V_1 \\ V_2 \end{bmatrix} = \frac{1}{3}\begin{bmatrix} 1 & 1 & 1 \\ 1 & a & a^2 \\ 1 & a^2 & a \end{bmatrix}\begin{bmatrix} V_a \\ V_b \\ V_c \end{bmatrix}, \text{ where } a = e^{j120^\circ} \tag{2.21}$$

where, the positive-sequence component (V_1) correspond to a balanced three-phase set with normal phase rotation (ABC), dominating under steady-state operating conditions and operating as the primary source of electromagnetic torque in electrical machines. The negative-sequence component (V_2) is a balanced set but with reverse phase rotation (ACB), which appears during unbalanced disturbances, faults, or asymmetries, and produces double-frequency currents in rotating machines that cause heating and mechanical stress. The zero-sequence component (V_0) consists of three phasors equal in magnitude and phase, and its behavior depends strongly on system grounding and transformer winding configuration. The inverse transformation recreates the phase quantities from the sequence components as:

$$\begin{bmatrix} V_a \\ V_b \\ V_c \end{bmatrix} = \begin{bmatrix} 1 & 1 & 1 \\ 1 & a^2 & a \\ 1 & a & a^2 \end{bmatrix} \begin{bmatrix} V_0 \\ V_1 \\ V_2 \end{bmatrix} \tag{2.22}$$

By decoupling the system into three independent sequence networks, sequence impedances Z_0, Z_1, and Z_2, faults such as single-line-to-ground, line-to-line, and double-line-to-ground can be evaluated across much simpler equivalent circuits.

Symmetrical components, therefore, provide a unified and highly efficient framework for studying unbalance, determining fault currents, analyzing machine heating due to negative-sequence currents, and ensuring proper protection coordination in both traditional rotating-machine-based systems and grids.

2.4 Basics of PowerFactory

The DIgSILENT GmbH, Gomaringen near Tübingen, is a software company providing highly specialized services in the field of electrical power systems for transmission, distribution, generation, industrial plants, and renewable energy. DIgSILENT's innovative product portfolio comprises PowerFactory, which is a leading power system analysis software, covering the full range of functionality from standard features to highly sophisticated and advanced applications. This software includes different toolboxes, namely, contingency analysis, Quasi-dynamic simulation, RMS/EMT, Arc flash analysis, cable analysis, economic analysis, and so on. In this section, the basics of this software, modelling, and load flow analysis of power systems by using PowerFactory will be covered. In this book, the PowerFactory 2024 has been used.

2.4.1 Getting Started

After the opening of PowerFactory, the initial step is to start the modelling of the power system. Initially, an empty window of PowerFactory can be seen, as shown in Fig. 2.1. To start the modeling of the power grid, click on File, then New > Project. Then, as shown in Fig. 2.2, a project overview window can be seen, and at the Basic Data tab, the name of the project can be added. Here, we added the name of the project as Test. After clicking on Ok, the grid overview window will be opened. In this window, you can name your grid and the frequency of this power system, as shown in Fig. 2.3. We used the name of Grid with a 50 Hz nominal frequency.

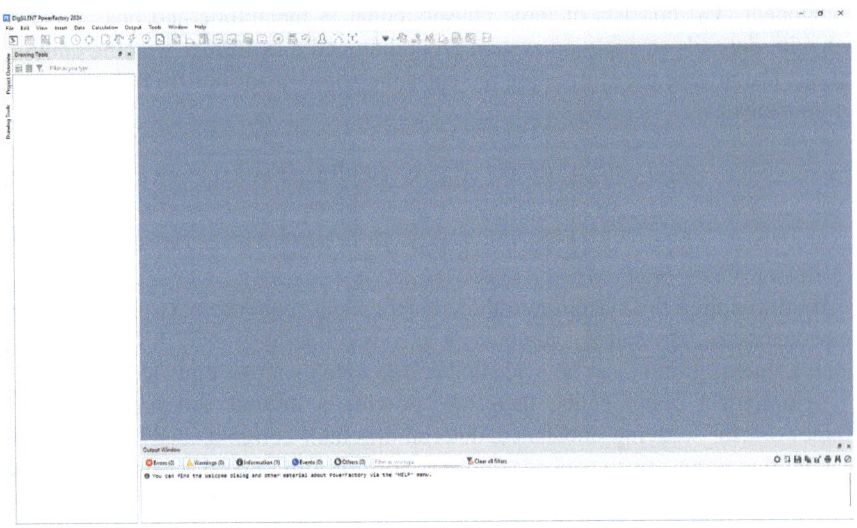

Fig. 2.1 PowerFactory window

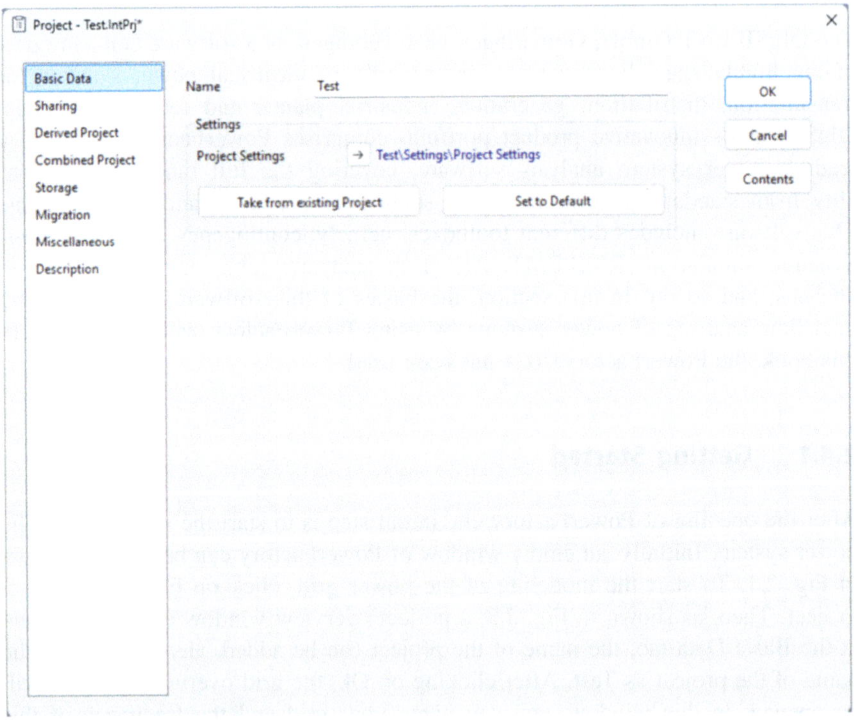

Fig. 2.2 New project window

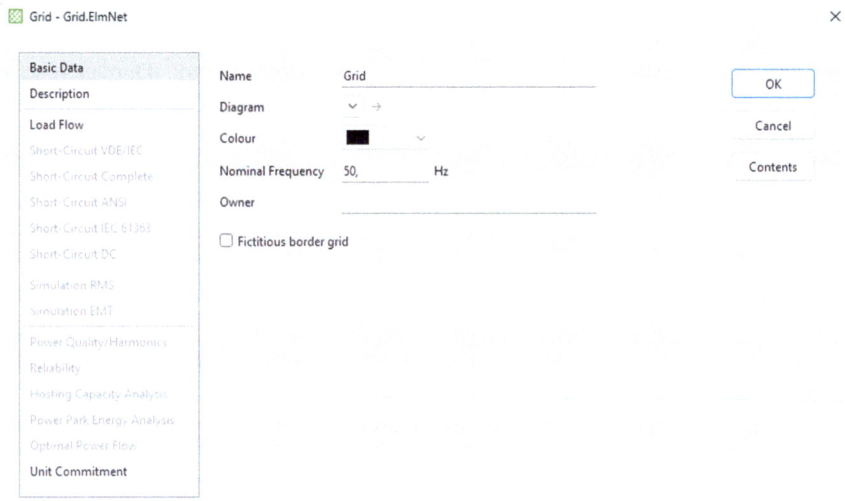

Fig. 2.3 New grid overview window

At this stage, a grid and a project have been defined, and the PowerFactory modeling environment is activated. As can be seen in Fig. 2.4, section 1 includes the drawing tools:

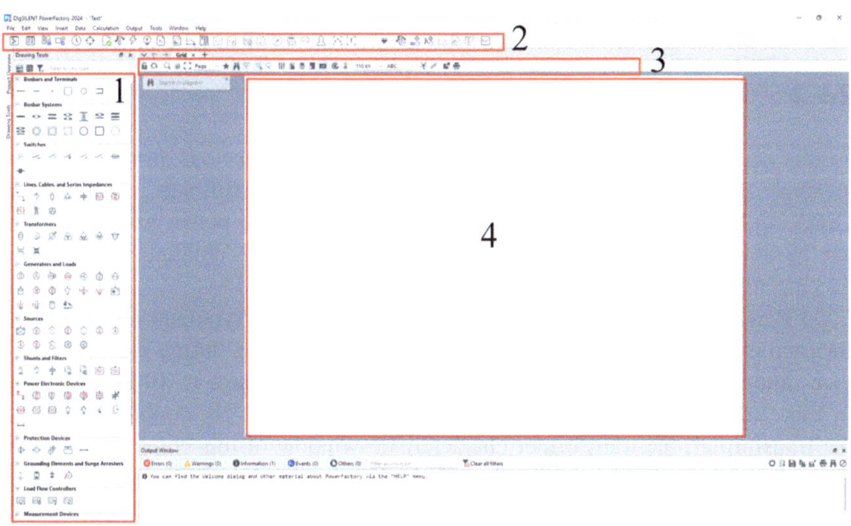

Fig. 2.4 PowerFactory main window including (1) drawing tools, (2) toolbox, (3) page layouts, (4) diagram

- Busbars and Terminals
- Switches: Circuit breakers and fuses
- Lines, Cables, and Series Impedances: Lines, Series reactor, Duplex reactor, Series capacitor, Series RLC-filter, Common Impedance, Fault current limiter, line coupling, cable system
- Transformers: 2-winding transformers, Auto-transformer, step-voltage regulator, 3-winding transformer, 4-winding transformer, multi-winding transformer, Booster transformer, Single phase booster transformer, and inductive coupling DC
- Generators and Loads: Synchronous and asynchronous machines, doubly fed induction machine, static generator, wind generator, battery, fuel cell, PV, DC machine, general load, MV load, LV load, connection request, DC load, PV panel, train
- Sources: external grid, AC voltage source, AC current source, DC voltage source, DC current source, impulse source,
- Shunt and Filters: RLC, RL, C, RLCR, RLCCR, and harmonic filters, static var system
- Power electronic devices: HVDC, rectifier, PWM converter, soft starter, TCSC, SSSC, DC/DC converter, DC valve, DC valve thyristor, DC valve IGBT, MMC valve, AC/DC converter

Each component can be selected from the section drawing tools and then added to the model by clicking on the desired point in section 4. Section 2 is the toolbox of the PowerFactory, which in this book each tool will be introduced in detail. Sections 3 and 4 are the settings of the diagrams and the grid diagram, respectively.

2.4.2 Power Systems Modelling

In this section, a step-by-step modelling and steady-state load flow analysis of a simple model is presented. By selecting two "Single Busbars", an "External Grid", and a "General Load" from the drawing tools, the power grid in Fig. 2.5 can be drawn. After installing the busbars by selecting "2-Winding Transformer" and installing it in the diagram, the model can be finalized.

In PowerFactory, after drawing the grid, inserting the type and characteristics for each element is a mandatory step to avoid further errors during analysis. To do that, double-click on the load busbar and change the voltage to 10 kV, as shown in Fig. 2.6. The grid's busbar voltage should be 110 kV. By double-clicking on the transformer, the transformer characteristic window of this component will be opened, as shown in Fig. 2.7. Then, by selecting the "New Project Type", the window of Fig. 2.8 will be opened, and the type of transformer should be defined

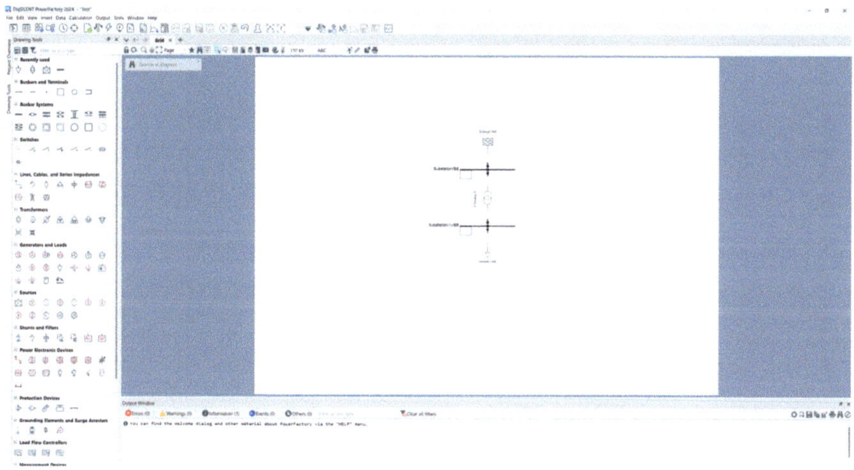

Fig. 2.5 A two-bus grid-connected system

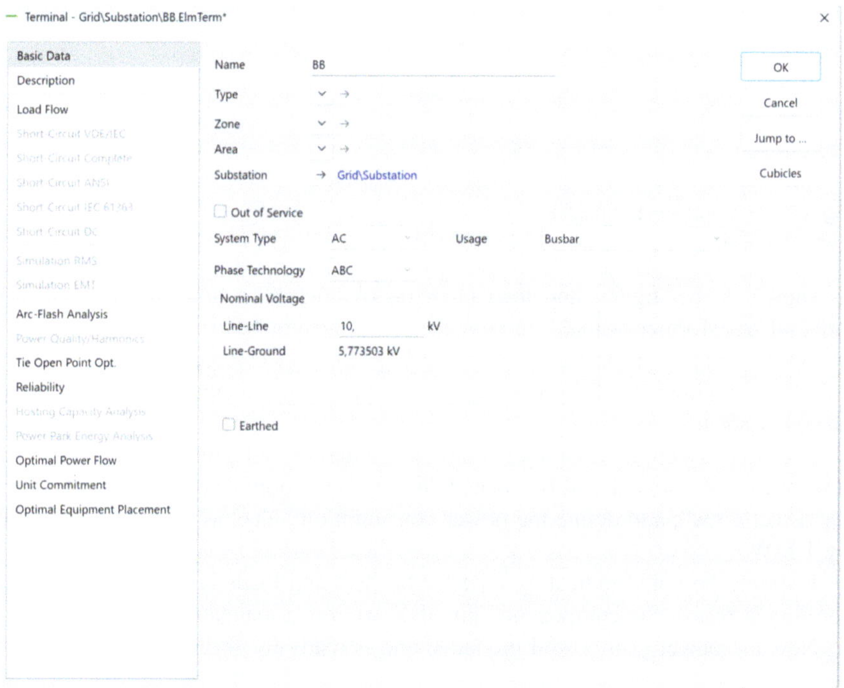

Fig. 2.6 Load busbar specifications

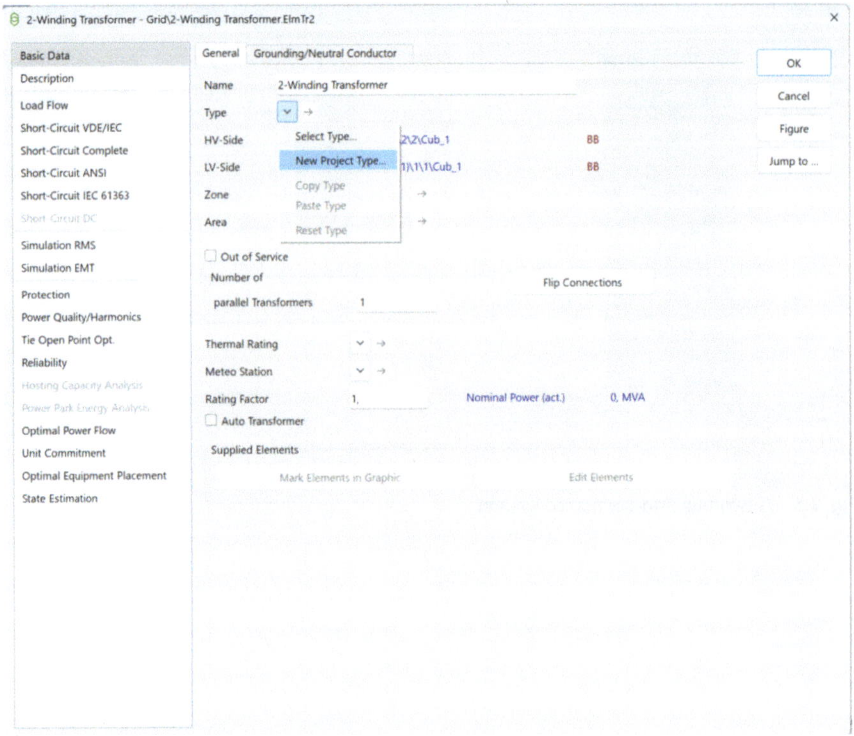

Fig. 2.7 Transformer specifications

as shown in this figure, and then click on ok. Just as a note, the LV and HV sides of transformers should be connected to the right busbar, which, during the incorrect connection, an error similar to Fig. 2.9 will emerge. This error can be solved by clicking the “Flip Connection” in the main window of the transformer characteristics.

For the last step, by double-clicking on “General Load”, the characteristic window of this component will be opened, and from the options on the left side, click on “Load Flow”, and define the power consumption value, as shown in Fig. 2.10, by 1 MW.

It should be noted indicating type for busbars and loads is not mandatory, and for almost rest of the components, this indication is a mandatory step.

Now the power grid is well modelled and is ready for further analysis.

2.4.3 Analysis of Three-Phase Power Systems

A load flow calculation provides voltages, currents, and powers at every point of a power grid. By clicking on the “Calculate Load Flow” in the toolbox section, as

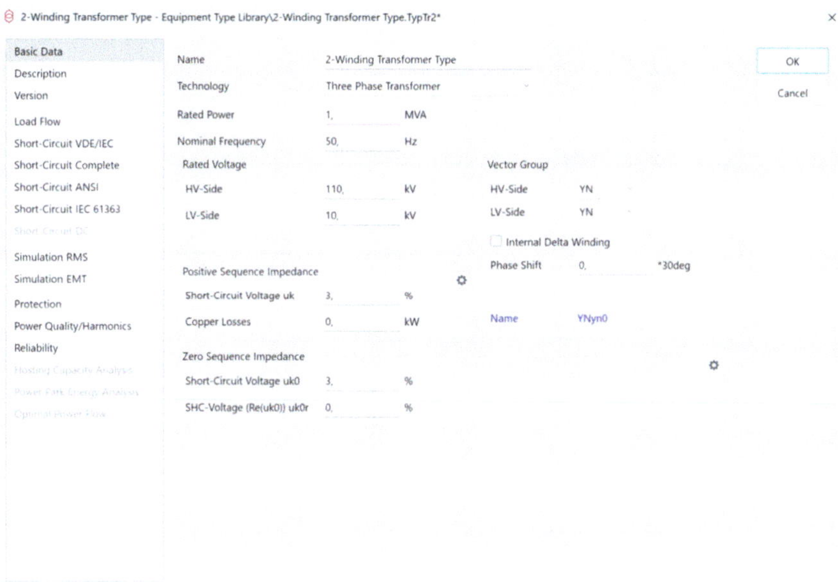

Fig. 2.8 Transformer type

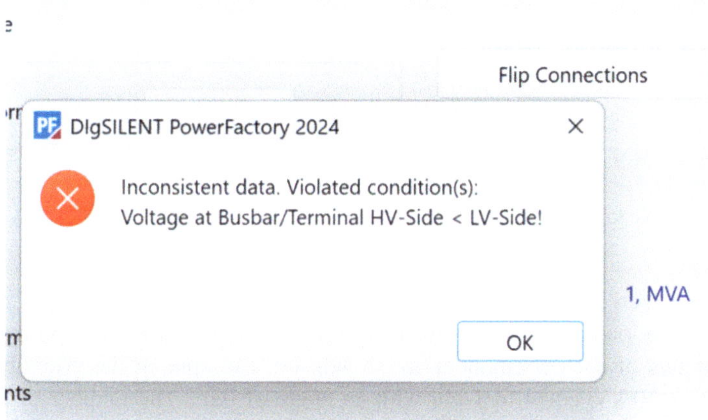

Fig. 2.9 Error during the incorrect connection of the transformer

shown in Fig. 2.11, the load flow toolbox will be opened, as shown in Fig. 2.12. As can be seen, PowerFactory calculates the load flow by three different methods: AC load flow balanced, AC load flow unbalanced, and DC load flow. For this example, use the AC load flow balanced and click on Execute to run the load flow analysis.

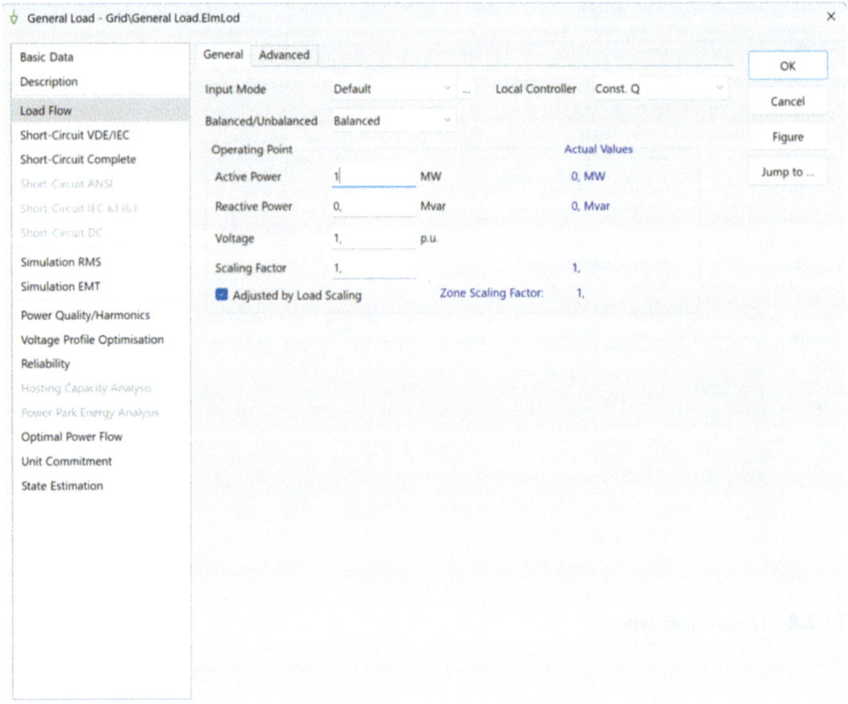

Fig. 2.10 General load specification

Fig. 2.11 Calculate load flow in the toolbox

As can be seen in the main diagram, the transformer color is changed to red, showing an overload at this component. By dragging the mouse scroll on the cube box, the results of each component are shown. For example, at the grid's busbar, the voltage is 10 kV, equal to 1 p.u with an angle of 0.0°. The number of results shown in this part is limited, and other result variables can easily be added to this part. By right-clicking on the results box and clicking on the "Edit Format for Nodes", as shown in Fig. 2.13, the detailed result window will be opened, as shown in Fig. 2.14.

By right-clicking on the empty space of the window and clicking on "Insert Row(s)", the result variable window will be opened, as shown in Fig. 2.15. As can be seen, as the last analysis applied to the system was Load flow AC, the calculation setting is selected to "Load Flow AC" by default. In the "Available variables" section, each required variable can be searched. Then, for observing the power factor, it should be selected.

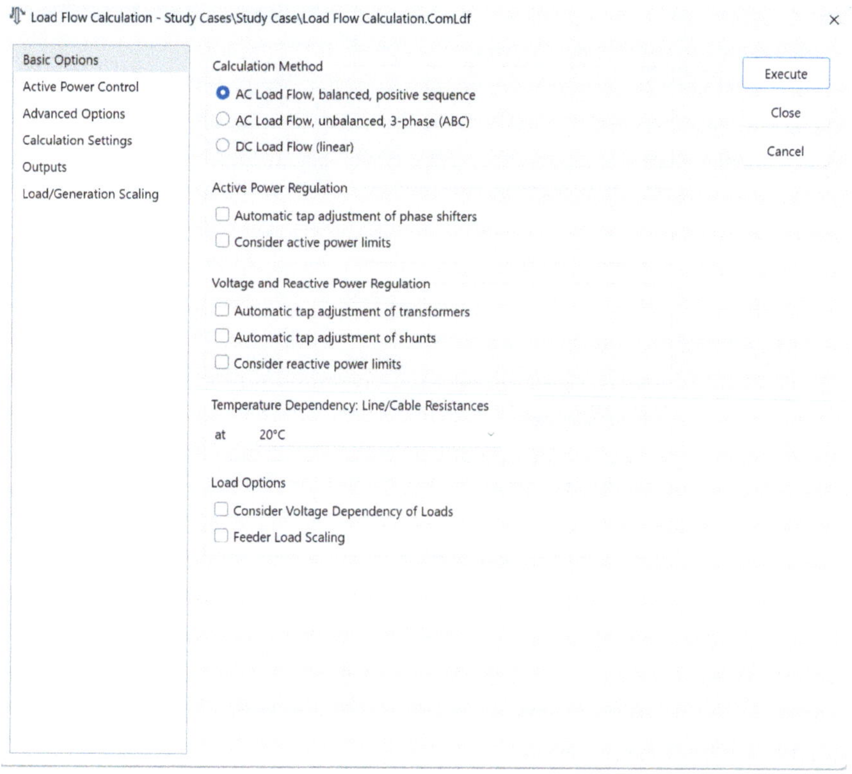

Fig. 2.12 Load flow toolbox

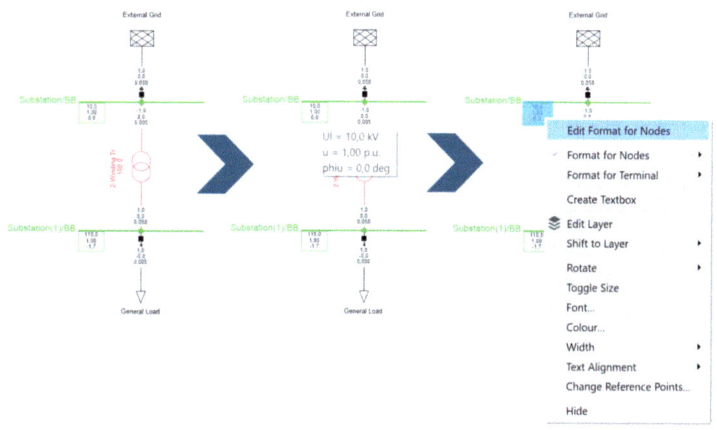

Fig. 2.13 Load flow output

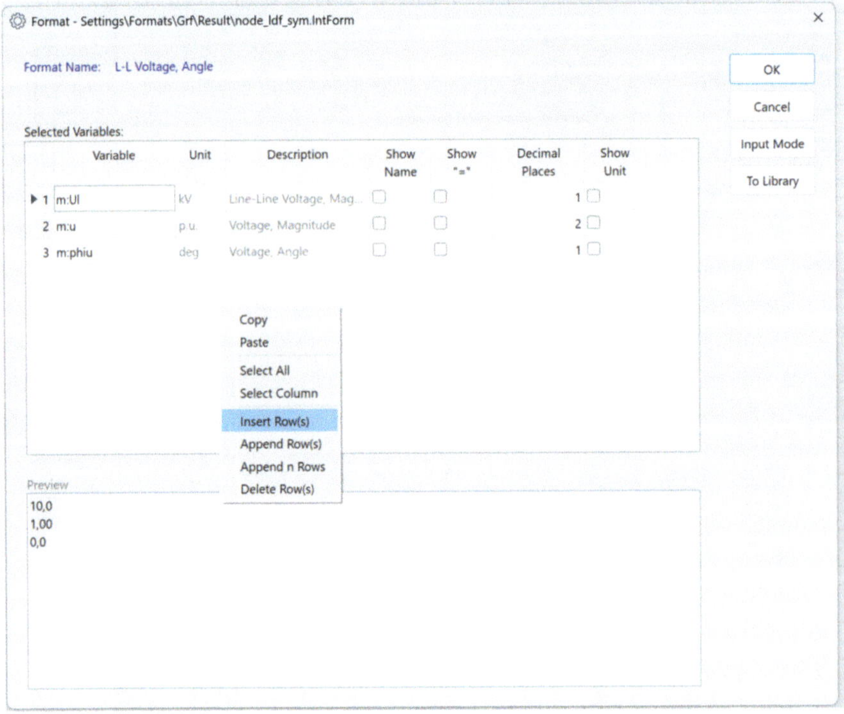

Fig. 2.14 Result box format window

As can be seen in Fig. 2.16, the results box is updated, and the power factor is observed with a value of 1,00, as there is no reactive power flow in the system.

PowerFactory includes several useful examples for further learning and observing different components. To open one of these examples, click on the File, then Examples to have access to the PowerFactory Example window as shown in Fig. 2.17. There are a couple of examples from real systems, in addition to the standard systems from CIGRE and IEEE. By clicking on the Transmission system example and importing it, this example will be opened, as shown in Fig. 2.18. For a practice, the load flow calculation can be performed in this example.

In PowerFactory, there is no save or load option, and all modeling and practices are saved automatically. To have access to previous models, you can click on File and then Activate project and find the desired project, and click ok.

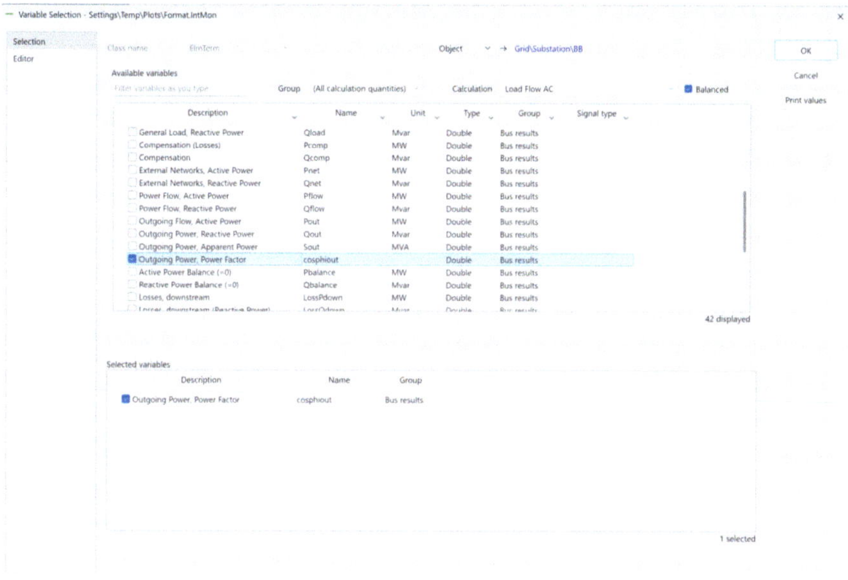

Fig. 2.15 Results variable window

2.5 Summary

This chapter has established the foundational principles of power system analysis by examining both single-phase and three-phase AC networks. In single-phase systems, the sinusoidal variation of voltage and current, RMS quantities, instantaneous power, and the decomposition into active and reactive power components were discussed in detail, along with the importance of power factor in system efficiency. The advantages of three-phase systems for bulk power transfer were then presented, including star and delta configurations, line-to-phase relationships, balanced power calculations, per-unit system normalization for simplifying multi-component network studies, and symmetrical components for handling unbalanced conditions and fault analysis. Finally, a practical introduction to PowerFactory—DIgSILENT was provided, demonstrating step-by-step modeling of a simple two-bus system, data entry for components such as transformers and loads, execution of balanced AC load-flow calculations, and visualization of results. These concepts and tools form the essential basis for subsequent chapters on power system components, load flow, short-circuit analysis, protection, and stability studies.

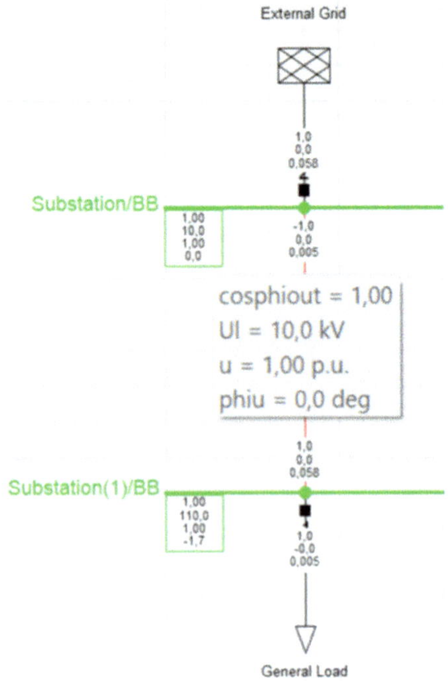

Fig. 2.16 Load flow results box, including power factor

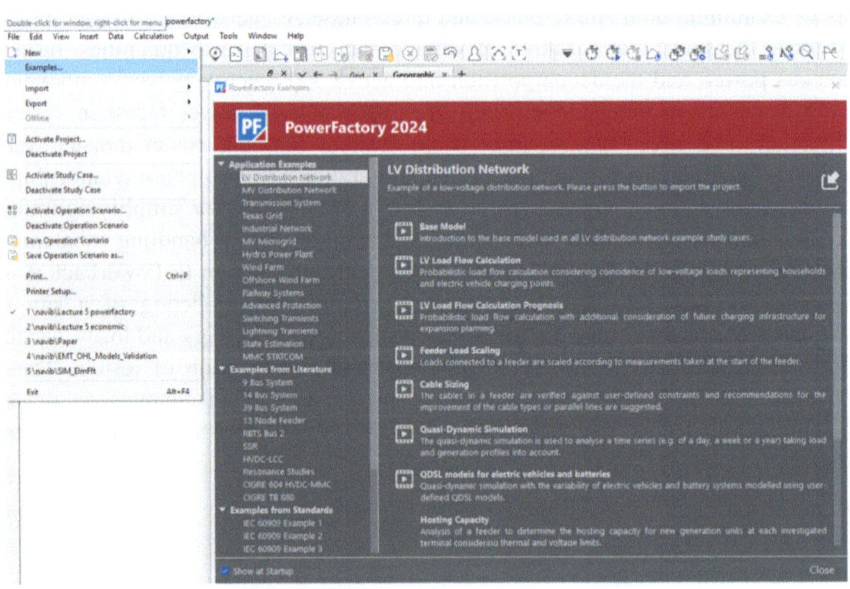

Fig. 2.17 PowerFactory Examples window

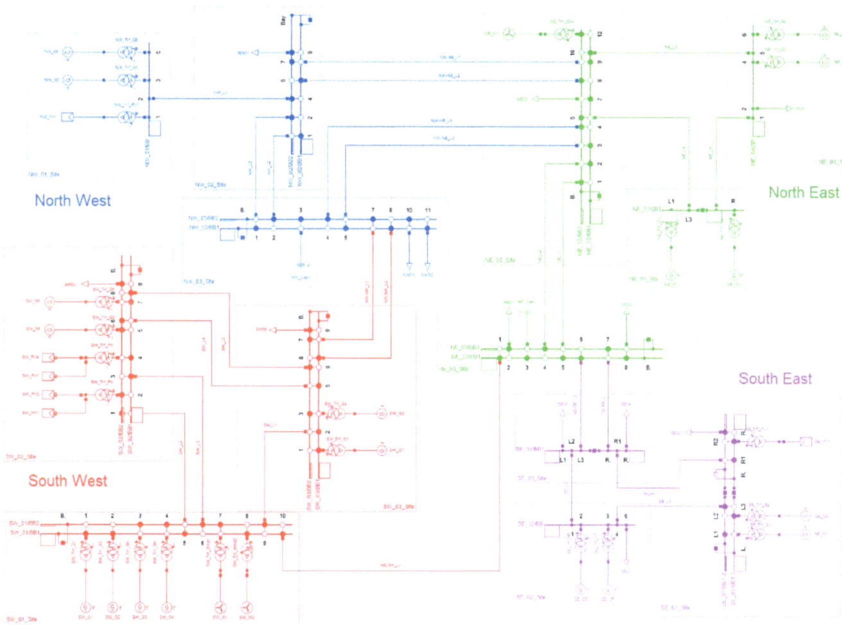

Fig. 2.18 Transmission system example of PowerFactory

2.6 Problems and Quest

1. Design the following power system in PowerFactory as shown in Fig. 2.19:

Transformer 132/33 kV: 90 MVA 132/33 kV (Grid 33 kV; line 132 kV).

Transformer 132/15.75 kV: 90 MVA 132/15.75 kV.

Line: 132 kV, R = 0.02, X = 0.11 Ohm/km, 50 km.

Load: S = 50 MVA, P.F = 1.

(a) Make a load flow and write down the voltages.
(b) Change the P.F. from 1 to 0.3. What happened to the voltages of busbars?
(c) Suggest and test 3 solutions for this problem.

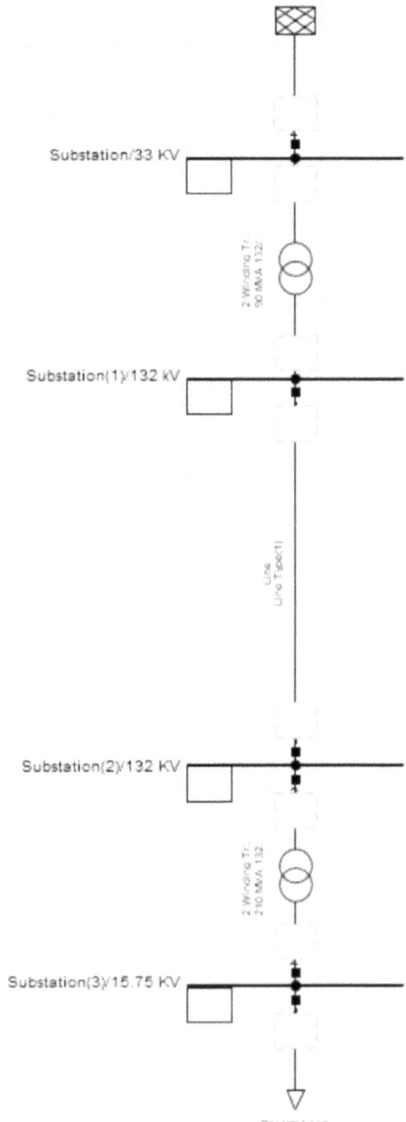

Fig. 2.19 Power system of practice 1

References

1. Grainger & Stevenson (1994) Power system analysis. Wiley
2. Elgerd (1982) Electric energy systems theory: an introduction. McGraw-Hill
3. Glover et al (2012) Power system analysis and design. Cengage/Elsevier

Transformers and Synchronous Machines

3

Abstract

This chapter has presented the fundamental theory, modeling, and practical aspects of two cornerstone components of power systems: transformers and synchronous machines. It began with the critical role of transformers in stepping up or stepping down voltage levels to enable efficient long-distance transmission, reduce I^2R losses, provide electrical isolation, supply neutral points in polyphase systems, and support accurate measurement through current and potential transformers. The ideal transformer model was introduced, covering voltage and current transformation ratios, power invariance, and impedance reflection. This was followed by the development of a realistic equivalent circuit that accounts for winding resistances, leakage reactances, core losses, and the magnetizing branch, enabling precise analysis of efficiency, voltage regulation, and performance under load. The chapter then extended to the working principles and detailed modeling of synchronous machines, from basic construction and operating principles to accurate equivalent-circuit representations suitable for steady-state and dynamic power-system studies. Finally, step-by-step modeling tutorials in PowerFactory—DIgSILENT were provided for both transformers and synchronous machines to reinforce theoretical concepts with practical simulation skills. The knowledge gained in this chapter forms an essential foundation for subsequent chapters on power flow, short-circuit analysis, protection, and stability studies.

Keywords

Transformers • Equivalent circuit • Synchronous machine

N. Bayati and M. A. Khan, *Power System Fundamentals with PowerFactory*,
https://doi.org/10.1007/978-3-032-32450-4_3

3.1 Introduction

In power systems, the majority of power should be transferred from far power plants to consumers like factories and homes through transmission lines. However, there are two obstacles to efficient power transfer in power grids: line capacity and losses. The main parameter that could have an impact on these two obstacles is current, as a higher current flow means higher losses, and the line capacity is limited by the amount of current the conductors can handle. As the amount of power is constant, due to the need of electrical consumers, by increasing the voltage level, the current level can be reduced. In power systems, transformers are passive components that can increase or decrease voltage levels, and consequently, the current level also will change.

Transformers are used to change only AC voltage levels, such as step-up or step-down types, to increase or decrease the voltage level, respectively, depending on where transformers are located, as shown in Fig. 3.1. Transformers were invented in 1885, and then they became essential for the transmission, distribution, and utilization of alternating current electric power. Transformers range in size from less than a cubic centimeter in volume to units weighing hundreds of tons used to interconnect the power grid.

In 2020, 4.4 million power transformers were in use in Europe, where 63% are transformers in distribution grids to power houses, 32% for distribution to industries, 3% to attach distributed energy resources like wind, solar, and hydro to the grid, and 2% in high-voltage transmission systems [1].

One of the main advantages of transformers in power grids is the ability to decrease losses during transmission and distribution of electrical power. Increasing voltage level by stepping up transformers at the generation side, the current flowing through the transmission lines will reduce for a given power level. This current reduction directly minimizes the power losses, RI^2, where I represents current and R is the resistance of the conductors. In HV transmission lines, such as those operating at 220 kV or higher, this efficiency gain is crucial, allowing utilities to transmit electricity over long distances with minimal energy losses as heat.

The power transfer capacity of transformers can be enhanced by enabling voltage transformation to higher levels, which permits higher power through existing infrastructure. The maximum power transfer capacity over a transmission line is governed by the thermal limits of the conductors. By increasing the voltage level,

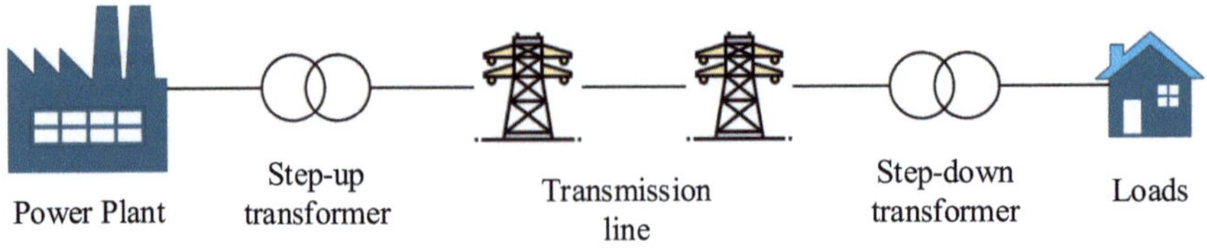

Fig. 3.1 Transformer applications in power systems

the system can handle higher power loads without exceeding current ratings or requiring thicker cables. Thus, it could optimize the utilization of grid assets. This is particularly vital in interconnected grids where load centers are far from generation sources.

Electrical isolation provided by transformers is another key benefit, ensuring the galvanic separation between the primary and secondary windings through the magnetic coupling of the core. This isolation prevents direct electrical connection and can protect sensitive equipment and personnel from HV faults or ground loops. In applications like distribution systems, it mitigates the propagation of transients, surges, or electromagnetic interference, improving system reliability and safety compliance. Moreover, in scenarios involving different grounding schemes or impedance matching, the isolating feature of transformers allows for independent operation of circuits, reducing the risk of cascading failures and enabling safer maintenance procedures. This is the reason that sometimes grid operators even use the transformers with a transformer ratio of 1.

Three-phase transformers provide a neutral point in polyphase systems, which is essential for grounding and balanced operation. In wye-connected transformers, the center point of the windings serves as the neutral point, which is essential for the connection of single-phase loads and providing a return path for unbalanced currents. This configuration supports effective grounding to limit overvoltages during faults and enables the use of ground fault protection relays. Without this neutral, systems would be prone to floating potential and increased stress on insulation.

In power systems, measuring voltage and current cannot be performed using the typical measurement units directly in the grid. Transformers serve as critical measurement units in power systems through specialized types like current transformers (CTs) and potential transformers (PTs), which step down high currents and voltages to safe and measurable levels for instrumentation. CTs operate on the principle of ampere-turn balance and provide a secondary current proportional to the primary (typically 5A or 1A), while PTs similarly reduce voltages (e.g., to 110 V) for accurate metering. These instrument transformers ensure precise monitoring of parameters like power factors, energy consumption, and fault currents.

3.2 Ideal Transformers

The varying AC current in the coil of transformers produces a varying magnetic flux in the transformer's core, and consequently, it induces a varying electromotive force (EMF) across the other coil of the same transformer.

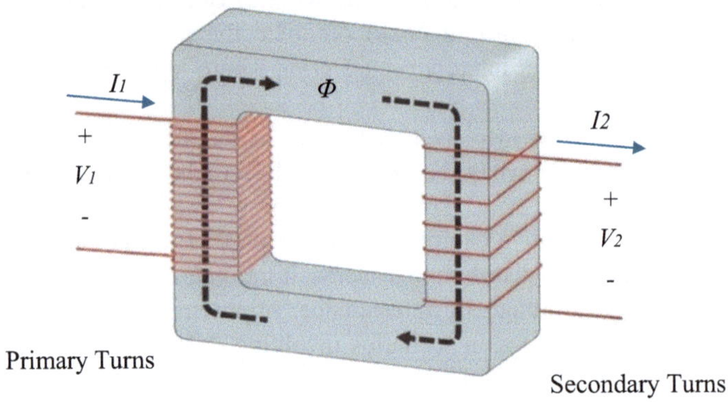

Fig. 3.2 Transformer winding and electrical parameters

As shown in Fig. 3.2, the primary winding has N_1 turns and is supplied with a voltage of V_1 and a current of I_1. Then, it produces a voltage of V_2 and current of I_2 at the secondary winding with N_2 turns. The current from the primary winding produces the common magnetic flux $\boldsymbol{\Phi}$ circulates through the core. In sinusoidal steady-state operation, Faraday's law gives the induced voltages as:

$$\begin{aligned} V_1 &= N_1 \frac{d\Phi}{dt} = j\omega N_1 \Phi \\ V_2 &= N_2 \frac{d\Phi}{dt} = j\omega N_2 \Phi \end{aligned} \tag{3.1}$$

Dividing these expressions yields the fundamental voltage transformation ratio:

$$\frac{V_2}{V_1} = \frac{N_2}{N_1} \tag{3.2}$$

Therefore, in ideal transformers, the delivered voltage can be calculated by knowing the number of turns and the primary voltage. For example, for a 1000:250 transformer with a primary voltage of 200 V, the secondary voltage will be 50 V.

In ideal transformers, the power delivered from primary to secondary is constant without any losses; therefore:

$$V_1 I_1 = V_2 I_2 \Rightarrow \frac{I_2}{I_1} = \frac{N_1}{N_2} \tag{3.3}$$

This relationship, combined with the voltage ratio, guarantees perfect power conservation.

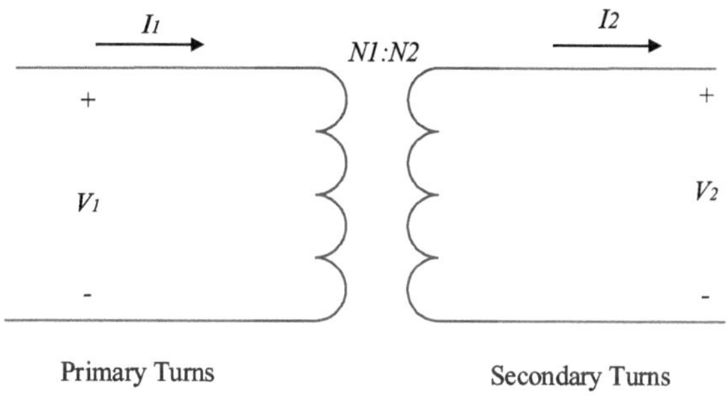

Fig. 3.3 Transformers in circuit schematics

On the other hand, when any loads are connected to the secondary of transformers, an impedance measurement from the primary side shows different impedance values. If an impedance of Z_2 is connected to the secondary winding, the measured impedance from the primary winding is

$$Z'_2 = \frac{V_1}{I_1} = \left(\frac{N_1}{N_2}\right)^2 Z_2 \tag{3.4}$$

In circuit schematic, transformers can be shown as Fig. 3.3, including two windings, and the winding turns are also shown as N1:N2.

Example 1 If the primary winding turns of a transformer have 2000 turns, and the secondary winding turns is 500, and $V_1 = 1200\measuredangle 0\,\mathrm{V}$ and $I_1 = 5\measuredangle -30\,\mathrm{A}$. Define the connected load at the secondary side.

Voltage transformation (secondary voltage):

$$V_2 = \frac{500}{2000} \times 1200\angle 0^\circ = 300\angle 0^\circ\,\mathrm{V}$$

Current transformation (secondary current):

$$I_2 = \frac{2000}{500} \times 5\angle -30^\circ = 20\angle -30^\circ\,\mathrm{A}$$

Load impedance (referred to secondary side):

$$Z_2 = \frac{V_2}{I_2} = \frac{300\angle 0^\circ}{20\angle -30^\circ} = 15\angle 30^\circ\,\Omega$$

3.3 Equivalent Circuit of Transformers

In power system studies, the equivalent circuit of transformers serves as the analytical method for understanding the behavior of single-phase and three-phase transformers under steady-state conditions. This model transforms the physical complexity of a transformer, including core and two coils, into a circuit of lumped impedances to capture the essential electrical phenomena such as ohmic losses in the windings, leakage fluxes that do not fully couple between primary and secondary, and the core's magnetizing requirements along with its associated energy dissipation, as shown in Fig. 3.4.

Referring to the primary side for convenience, the exact equivalent circuit positions the shunt branch, representing core effects, and two series RL circuits as the winding losses and leakage paths, as shown in Fig. 3.5.

Here, the primary voltage V_1 drives the input current I_1 through the primary resistance r_1 and leakage reactance x_1, where $x_1 = \omega L_1$, with $\omega = 2\pi f$, the angular frequency. Then, current reaches the parallel combination of the core-loss conductance G, and magnetizing susceptance B ($1/X_m$, with $X_m = \omega L_m$). This shunt branch draws the excitation current, which sustains the mutual flux while accounting for core losses. Then, current reaches the secondary resistance $a^2 r_2$ and leakage reactance $a^2 x_2$, where a is the transform ratio, and it is squared due to moving the values from secondary to primary. The circuit then feeds an ideal transformer with a turns ratio a:1 ($a = N_1/N_2$), leading to the load current I_2 and voltage V_2.

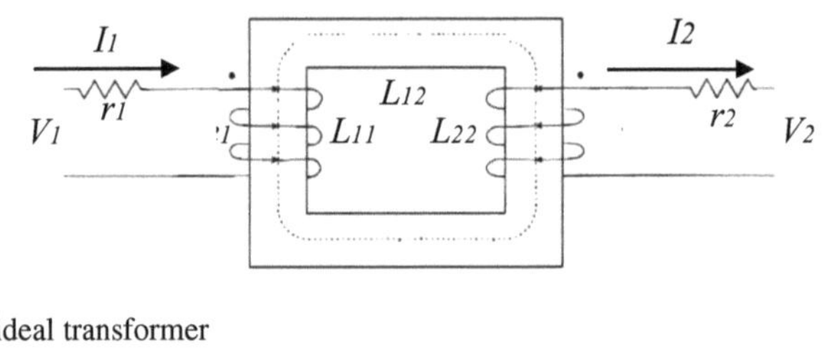

Fig. 3.4 Non-ideal transformer

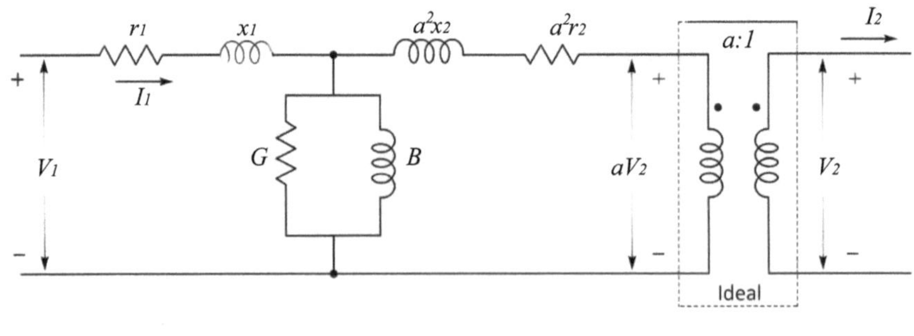

Fig. 3.5 Equivalent circuit of single-phase transformers

The primary terminal voltage decomposes as:

$$V_1 = I_1(r_1 + jx_1) + V_c \tag{3.5}$$

where V_c is the voltage across the core branch at the primary side. The no-load excitation current, comprising the in-phase core-loss component I_c and quadrature magnetizing component I_m, is given by:

$$I_0 = V_c(G + jB) = V_c\left(\frac{1}{R_c} + j\frac{1}{X_m}\right) \tag{3.6}$$

or, in magnitude form:

$$I_0 = \frac{V_c}{\sqrt{R_c^2 + X_m^2}} \tag{3.7}$$

On the secondary side, referred to the primary, the voltage equation becomes:

$$aV_2 = (I_1 - I_0)\left(a^2 r_2 + ja^2 x_2\right) - V_c \tag{3.8}$$

These relations highlight the transformer's non-ideal nature, that the primary current exceeds the referred load current due to the small but non-negligible core leakage current, typically 2–5% of full-load current, and voltage drops occur across both series branches.

Therefore, it is obvious that in no-load conditions, the secondary voltage is maximum, and in full load conditions, due to the higher current flow, the secondary voltage is minimum. Voltage regulation in transformers is the percentage change in secondary terminal voltage from no-load to full-load, relative to the no-load voltage, while keeping primary supply voltage constant. It measures a transformer's ability to maintain a stable, near-constant output voltage, with lower percentages indicating more stable performance. Voltage regulation, a key performance metric quantifying the change from no-load to full-load voltage, is expressed as:

$$VR = \frac{V_{2,no\text{-}load} - V_{2,full\text{-}load}}{V_{2,full\text{-}load}} \times 100\% \tag{3.9}$$

where, $V_{2,no\text{-}load}$ and $V_{2,full\text{-}load}$ are the no-load and full-load voltages at the secondary, respectively.

Example 2 A single-phase transformer has 1000 turns on the primary side and 250 turns on the secondary. Winding resistances of the primary side are 2 Ω, and the secondary side is 0.125 Ω. Leakage reactances are $x_1 = 8\ \Omega$, and $x_2 = 0.5\ \Omega$. The resistance load is 12 Ω. If the applied voltage at the terminals of the primary winding is 1000 V, find the V2 and the voltage regulation. Neglect magnetizing current.

$$a = \frac{N_1}{N_2} = \frac{1000}{250} = 4$$
$$R_1 = 2 + 0.125(4)^2 = 4\,\Omega$$
$$X_1 = 8 + 0.5(4)^2 = 16\,\Omega$$
$$Z'_2 = 12(4)^2 = 192\,\Omega$$
$$I_1 = \frac{1000\measuredangle 0}{192+4+j16} = 5.08\measuredangle -4.67\text{ A}$$
$$aV_2 = 5.08\measuredangle -4.67 \times 192 = 976\measuredangle -4.67\text{ V}$$
$$V_2 = \frac{976\measuredangle -4.67}{4} = 244\measuredangle -4.67\text{ V}$$

$$\text{Voltage Regulation} = \frac{1000/4 - 244}{244} = 0.0245\text{ or }2.45\%$$

3.3.1 Short-Circuit and Open-Circuit Tests

The equivalent circuit of a single-phase transformer models its electrical behavior by representing the primary and secondary windings as series impedances (accounting for winding resistances and leakage reactances) connected through an ideal transformer, with a shunt branch across the primary to model core losses and magnetizing effects. The parameters include primary resistance R_1, secondary resistance R_2, primary leakage reactance X_1, secondary leakage reactance X_2, core loss resistance R_c, and magnetizing reactance X_m. These are typically referred to one side (e.g., primary) as equivalent resistance $R_{eq} = R_1 + a^2R_2$, equivalent leakage reactance $X_{eq} = X_1 + a^2X_2$, where $a = N_1/N_2$ is the turns ratio. The open circuit and short circuit tests provide empirical data to determine these parameters without disassembling the transformer, assuming sinusoidal steady-state operation at rated frequency.

The open circuit test is conducted by applying the rated voltage V_1 to the primary winding while keeping the secondary open-circuited, typically on the low-voltage side to minimize instrumentation requirements. Under these conditions, the input current I_{oc} is primarily the no-load current (magnetizing plus core loss components), and the input power P_{oc} equals the core losses (hysteresis and eddy current). The measured quantities are V_{oc}, I_{oc}, and P_{oc}. The core loss resistance R_c is calculated as $R_c = V_{oc}^2/P_{oc}$, representing the equivalent resistance dissipating core losses. The magnetizing reactance X_m is derived from the no-load admittance $Y_{oc} = I_{oc}/V_{oc}$, where the conductance $G_c = 1/R_c$ and susceptance $B_m = 1/X_m$, satisfying $Y_{oc} = \sqrt{G_c^2 + B_m^2}$. Thus, $X_m = V_{oc}/\sqrt{I_{oc}^2 - (P_{oc}/V_{oc})^2}$. This test neglects winding losses since I_{oc} is small (typically 2–5% of rated current), isolating the shunt branch parameters.

$$\begin{aligned} P &= \frac{V_2}{R_m} \\ Q &= \sqrt{(V_1I_1)^2 - P^2} = \frac{V_1^2}{X_m} \end{aligned} \tag{3.10}$$

The short circuit test involves shorting the secondary terminals and applying a reduced voltage V_{sc} to the primary (usually on the high-voltage side) until rated current I_{sc} flows, ensuring full-load copper losses without excessive heating. The measured quantities are V_{sc}, I_{sc}, and P_{sc}. Under short circuit, the core flux is negligible due to opposing MMFs, so the input power P_{sc} approximates total copper losses $I_{sc}^2 R_{eq}$, yielding $R_{eq} = P_{sc}/I_{sc}^2$. The equivalent impedance $Z_{eq} = V_{sc}/I_{sc}$, and thus the equivalent leakage reactance $X_{eq} = \sqrt{Z_{eq}^2 - R_{eq}^2}$. If referred to the primary, these directly give the series branch; otherwise, scaling by a^2 refers them to the secondary

$$\begin{gathered} R_e = \frac{P}{I_1^2} \\ Q = \sqrt{(V_1 I_1)^2 - P^2} = X_e I_1^2 \end{gathered} \tag{3.11}$$

Combined, these tests enable the full equivalent circuit: the open circuit provides R_c and X_m in parallel across the primary, while the short circuit provides R_{eq} and X_{eq} in series before the ideal transformer. Efficiency and voltage regulation can then be computed, e.g., efficiency $\eta = (V_2 I_2 \cos\phi)/(V_2 I_2 \cos\phi + P_{oc} + P_{sc})$, where $\cos\phi$ is the power factor. The connection of each test is shown in Fig. 3.6.

Example 3 An open-circuit test on a transformer produced the following measurements: $V_1 = 100$ V, $V_2 = 50$ V, $P = 9.5$ W, and $I_1 = 180$ mA. Determine the transformer turns ratio and the reactance and resistance of the core.

Turns ratio (from open-circuit test voltages or nameplate values):

$$a = \frac{V_{1,\mathrm{LV}}}{V_{1,\mathrm{HV}}} = \frac{50}{100} = 0.5$$

Core loss resistance (referred to the low-voltage side, from open-circuit test):

$$R_m = R_c = \frac{V_1^2}{P_{oc}} = \frac{100^2}{P_{oc}} \Rightarrow R_m = 1.05\,\mathrm{k}\,\Omega$$

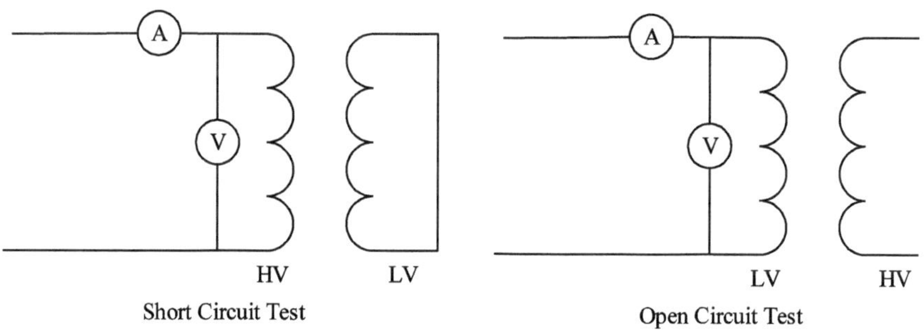

Fig. 3.6 Short-circuit and open-circuit tests

Magnetizing reactance calculation (from open-circuit test data):
First, the reactive power component (or magnetizing VA) is found using:

$$Q = \sqrt{(V_1 \cdot I_{oc})^2 - P_{oc}^2} = \sqrt{(100 \times 0.180)^2 - 9.5^2} = 15.26\,var$$

Then the magnetizing reactance (referred to the same side) is:

$$X_m = \frac{V_1^2}{Q} = \frac{100^2}{\sqrt{(100 \times 0.180)^2 - 9.5^2}} = 655\ \Omega$$

3.4 Three-Phase Transformers

Unlike single-phase transformers, a three-phase transformer is made by three windings for the primary and three windings for the secondary side, each side of which has the same characteristics to provide three sinusoidal balanced voltages and currents, and delivering balanced active power, $P = 3V_{ph}\ I_{ph}\ \cos\varphi$. The transformer consists of three sets of primary and secondary windings mounted on a shared magnetic core, typically a three-limb core-type construction using high-permeability grain-oriented silicon steel laminations to minimize eddy-current and hysteresis losses, as shown in Fig. 3.7.

Each phase operates with the same turns ratio, $a = N_1/N_2$, and inducing phase voltages according to Faraday's law, the same as the single-phase transformers. In three-phase transformers, we have two different definitions: phase and line.

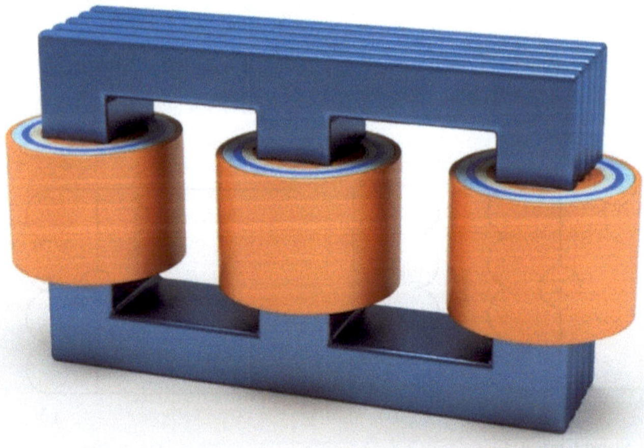

Fig. 3.7 Three-limb core-type of a three-phase transformer

Phase is the winding of each phase of a transformer, for example, phase voltage or current, meaning the voltage or current across each winding inside the transformer. While the line voltage or current means the voltage or current across the output line of the transformer in outside of it. Depending on the configuration of each transformer's side, we should have different considerations. Figure 3.8 shows diagrams for the four common three-phase transformer connections, illustrating winding arrangements, voltage/current phasors, and phase shifts. These are based on standard configurations for step-down operation.

1. Wye-Wye (Y-Y) Connection, Fig. 3.8a

Y-Y connections maintain 0° phase shift, with neutrals accessible for grounding. Line voltages are $\sqrt{3}$ times phase voltages, $V_L = \sqrt{3}V_{ph}$, and phase and line currents are equal, $I_L = I_{ph}$. Prone to zero-sequence flux issues in core-type transformers; often use tertiary delta for stability, and also as it can provide neutral points, this connection can be used in distribution grids.

2. Delta-Delta (Δ-Δ) Connection: Fig. 3.8b

In Δ-Δ, both primary and secondary are delta-connected, providing no phase shift (0°). Line voltages equal phase voltages on both sides, $V_{L,p} = V_{ph,p}$, $V_{L,s} = V_{ph,s}$, and line currents are $\sqrt{3}$ times phase currents. Suitable for balanced loads; circulating currents handle triple harmonics.

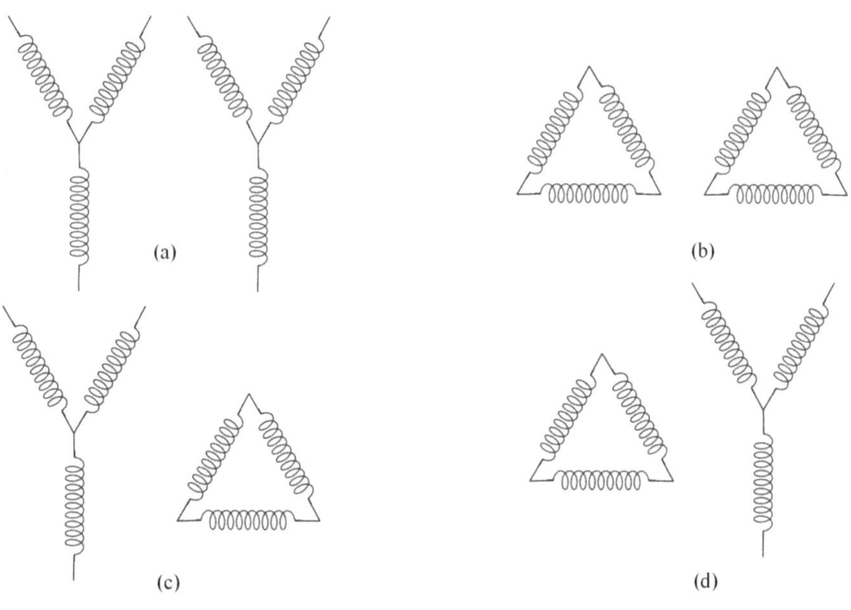

Fig. 3.8 Different configurations of three phase transformers **a** Y-Y **b** Delta-Delta **c** Y-Delta **d** Delta-Y

3. Wye-Delta (Y-Δ) Connection: Fig. 3.8c

Y-Δ has a 30° lead (secondary leads primary). Useful for step-down applications; wye primary allows grounding, delta secondary isolates harmonics. The Y side can be connected to the higher voltage, which provides less voltage across phases, resulting in lower insulation costs. And Δ will be connected to the LV side with a higher current, which means the current in each phase will be less, concluding less wiring and heating issues.

4. Delta-Wye (Δ-Y) Connection: Fig. 3.8d

Δ-Y introduces a 30° phase shift (secondary lags primary). Primary line voltage = phase voltage, secondary line voltage = $\sqrt{3}$ phase voltage. Common for step-up applications.

Here are the key equations for calculations of three-phase transformers:

1. Voltage Relationships:
   - Wye (Y): $V_L = \sqrt{3}V_{ph}$, $I_L = I_{ph}$
   - Delta (Δ): $V_L = V_{ph}$, $I_L = \sqrt{3}I_{ph}$
   - Transformation (per phase): $V_{ph,s} = \frac{V_{ph,p}}{a}$, $I_{ph,s} = aI_{ph,p}$
2. Power (Three-Phase):
   - Apparent power: $S = \sqrt{3}V_L I_L$
   - Active power: $P = \sqrt{3}V_L I_L \cos\phi$
   - Reactive power: $Q = \sqrt{3}V_L I_L \sin\phi$
3. Phase Shift in Mixed Connections (Δ-Y example, secondary lags primary by 30°):
   - Line voltages: $\vec{V}_{AB,s} = \frac{\sqrt{3}V_{L,p}}{a}\angle -30°$
   - Line currents (balanced): $\vec{I}_A = \frac{I_{L,s}}{\sqrt{3}}\angle 30°$ (for Y secondary)

Example 4 A 60–10 kV transformer is shown in Fig. 3.9. The phase voltages from both primary and secondary sides are shown in Fig. 3.9.

Example 5 A three-phase Y-Δ transformer with a rating of 75 MVA, 66/6.6 kV, connected with a balanced load of three 0.6 Ω, Y-connected resistors. Determine the per-unit resistance of the load on the base for the low-voltage side. Then, determine the load resistance in ohms referred to the high-voltage side and the per-unit value of this resistance on the chosen base.

The base impedance on the LV side is:

$$Z_{\text{base,LV}} = \frac{V_{\text{base,LV}}^2}{S_{\text{base}}} = \frac{6.6^2}{75} = 0.5808\ \Omega$$

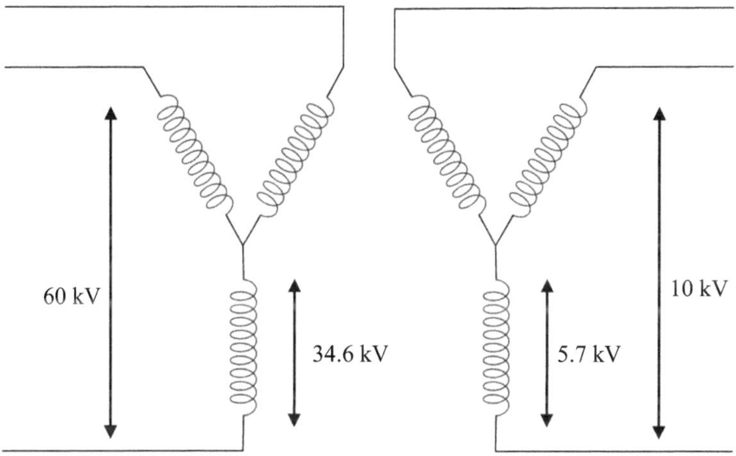

Fig. 3.9 Line and phase voltages of a Y-Y transformer

The per-unit resistance referred to the LV side (given R = 0.6 Ω on LV):

$$R_{\text{pu,LV}} = \frac{R_{\text{LV}}}{Z_{\text{base,LV}}} = \frac{0.6}{0.5808} = 1.03\,\text{pu}$$

The base impedance on the high-voltage (HV) side is:

$$Z_{\text{base,HV}} = \frac{V_{\text{base,HV}}^2}{S_{\text{base}}} = \frac{66^2}{75} = 58.1\ \Omega$$

The resistance referred to the HV side (using the square of the turns ratio):

$$R_{\text{HV}} = R_{\text{LV}} \times \left(\frac{V_{\text{HV}}}{V_{\text{LV}}}\right)^2 = 0.6 \times \left(\frac{66}{6.6}\right)^2 = 60\ \Omega$$

Finally, the per-unit resistance referred to the HV side is:

$$R_{\text{pu,HV}} = \frac{R_{\text{HV}}}{Z_{\text{base,HV}}} = \frac{60}{58.1} = 1.03\,\text{pu}$$

Note: The per-unit resistance is the same on both sides, as expected, because the transformation by the square of the turn's ratio cancels when converting between sides. Therefore, in studies, when per-unit calculation is used, the influence of transformers can be neglected.

3.4.1 Group Numbers

Three-phase transformers are classified into vector groups based on the phase shift, angular displacement, between the primary and secondary line voltages, expressed using clock-hour notation, where each hour represents 30°.

Group number 1 corresponds to zero phase shift from transformer configurations YY0, and DD0. In these configurations, the primary and secondary line-to-line voltages are in phase, meaning corresponding phases have zero angular difference. This alignment simplifies paralleling of transformers and is common in industrial applications or when no phase shift is desired. The absence of phase shift also means that in YnYn, zero-sequence voltages appear equally on both sides, and third-harmonic fluxes can circulate freely in DD0 windings to reduce neutral voltage distortion.

Group number 2 represents a 180° phase shift from transformer configurations YY6, and DD6, where the secondary line voltages are exactly opposite in phase to the primary. This phase shift effectively inverts the voltage waveform, making it useful in specific system designs requiring phase opposition, although it is less frequently used in modern distribution networks.

Group number 3 denotes a –30° lag phase shift on the secondary side, typical of DY1, and YD1connections. Conversely, group number 4 indicates a + 30° lead phase shift with DY11, and YD11 configurations, where the secondary leads the primary by 30°. These ±30° phase shifts arise from the vector sum of phase and line voltages in mixed delta-wye or wye-delta arrangements and are the most prevalent in distribution transformers, especially DYn11, because the delta winding traps third-harmonic currents, while the wye secondary provides a grounded neutral for unbalanced loads and single-phase consumers.

In summary:

Group number 1: No Phase displacement YY0, DD0.

Group number 2: 180° phase displacement YY6, DD6.

Group number 3: −30° displacement DY1, YD1.

Group number 4: +30° phase displacement DY11, YD11.

3.4.2 Tap-Changer

Tap changers are critical devices in power transformers that enable adjustment of the effective turns ratio to regulate the secondary voltage under varying primary voltage and load conditions. They are usually located on the HV winding near the neutral point to reduce insulation requirements during switching. On the HV side, the current is less, and during the changing of the taps, the possibility of having an arc is less than on the LV side. A tap changer including 8 taps is shown in Fig. 3.10, where the voltage of the secondary can change between 96 to 110%.

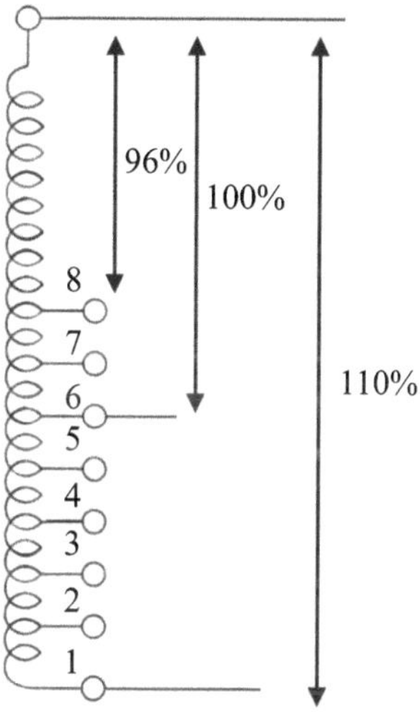

Fig. 3.10 A tap changer including 8 taps

The output voltage is directly determined by the turns ratio:

$$\frac{V_2}{V_1} = \frac{N_2}{N_1} \tag{3.12}$$

At a given tap position k (where k = 1 corresponds to the nominal tap), the actual turns ratio becomes:

$$a_k = \frac{a_0}{k} \tag{3.13}$$

A typical tap changer provides a regulation range of up to ±16% in small steps. Without tap adjustment, any deviation in primary voltage directly affects the secondary output:

$$V_2 \approx V_1 \times \frac{N_1}{N_2} \tag{3.14}$$

With tap compensation, the secondary voltage is maintained close to the rated value:

$$V_2 \approx V_{rated} \times \frac{k}{k_{nom}} \tag{3.15}$$

Automatic tap changer controllers usually employ a proportional-integral (PI) algorithm to decide the required tap movement.

Example 6 A 60/10 kV transformer has five taps on the primary winding, giving 2% voltage regulation by each tap, −4, −2%, nominal, 2, and 4%. If the voltage of the primary is reduced by 2%, then the secondary voltage at the load will be 9.8 kV. To raise secondary voltage from 9.8 to 10 kV, the number of turns on the primary side should be reduced by 2% to compensate for it. Therefore, the primary tap should be changed to −2% to provide regulation.

There are two types of tap-changing transformers:

1. Off-Load Tap-Changing Transformer: Typically, when the transformer is hot, changing the taps by moving the mechanical part provides an arc between taps. In more cost-effective solutions, an off-load tap-changer cannot remove these arcs or cause long-term damage to these transformers. Therefore, the tap for these transformers can be changed when transformers are not in operation, cold.
2. On-Load Tap-Changing Transformer: Unlike the off-load tap changers, the on-load tap changers use an oil tank and special switching between taps to avoid arc or remove it at the zero-crossing point very fast.

In different scenarios, tap changers should be used to improve the operation of grids. In urban distribution systems, residential and commercial loads fluctuate significantly between daytime peaks and nighttime demands. Therefore, during the peak hours, the voltage will drop, and during the low demands voltage will increase. Based on the regulations, voltage should be limited to within ±5% of nominal voltage. Without a tap changer, a sudden increase in inductive loads could cause a voltage drop exceeding 5%.

For grid-connected solar or wind farms, where primary voltage varies due to intermittent generation (e.g., cloud cover reducing PV output by 20–50%), tap changers compensate to keep substation voltages stable, avoiding reverse power flow or frequency instability.

In HV transmission over long lines, capacitive charging causes voltage rises or falls up to 15%, exacerbated by the Ferranti effect at light loads. Tap changers on step-down transformers at receiving ends adjust the ratio to reduce this voltage.

3.5 Synchronous Machines

Synchronous machines are electro-mechanical devices that convert the mechanical energy of the converter into electrical energy. It's widely used for power generation as well as for the regulation of power factors. It operates on the principle of rotating the rotor in synchronism with the rotating magnetic field, which is generated by the stator. This fixed speed of the motor under steady state conditions is necessary for obtaining a stable frequency and better voltage quality for interconnected power systems. Further, their ability to convert mechanical power into electrical power with high efficiency strengthens their prevalent position in utility-scale generation.

From an operational overview, synchronous machines are essential for large-scale electricity generation, as they convert mechanical power from wind or hydro-based turbines into electrical power and supply it to the grid. Their capability to control reactive power by adjusting excitation makes it essential for voltage

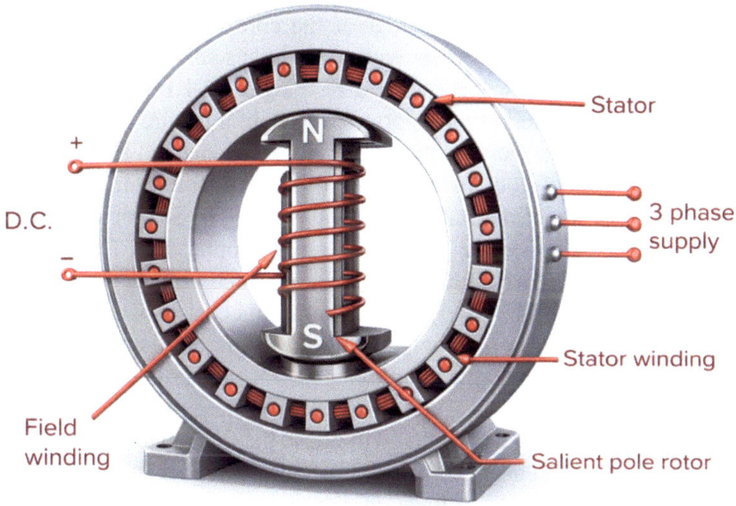

Fig. 3.11 Representative model for synchronous machine

regulation and reactive power management. In addition, synchronous machines can be used as synchronous condensers for providing reactive power compensation, while improving voltage stability, and backing grid inertia. Figure 3.11 shows the overall model of a synchronous machine.

Overall, the synchronous machine combines controllability, efficiency, and speed-based frequency control, which is critical for traditional and emerging power systems. With the increase in renewable energy sources, synchronous machines present necessary grid-support functions, making investigation vital for understanding electrical power engineering.

3.5.1 Synchronous Machines Structure

A synchronous machine consists of two main components: the stator and the rotor. The stator comprises a stationary armature that has a three-phase armature winding, whereas the armature has a DC excitation field. Torque is produced by the interaction of the rotating field of the stator and the rotor excitation field. The torque enables the operation at a constant speed. The speed is a correlation between electrical frequency and the number of poles, which can be denoted as:

$$n_s = \frac{120f}{P} \tag{3.16}$$

The stator is designed with a laminated steel core to reduce the eddy current losses. The slots along the inner border of the core tend to house the armature winding, which operates in 3 phases with a 120° phase displacement. Further, the stator assembly happened to be enclosed in a rigid steel frame, which provides mechanical strength as well as incorporates the cooling feature either through ventilation ducts or air vents. Further, the primary flux is generated by the rotor, which receives DC excitation through slip rings and brushes. The rotor contractions can be categorized into salient-pole and cylindrical rotors. The Salient pole architecture generally composes a projecting pole for low-speed application. They usually have a large number of poles with field coils wound around each pole and uneven air gaps between the poles. Moreover, the winding also consists of a damper winding on the faces of the poles, which helps damp down the oscillation and helps with starting for an asynchronous machine. whereas machines with cylindrical rotors are for high-speed applications. The construction of these rotors is done with solid steel, consisting of machined slots that will house the field windings, and a uniform air gap is maintained to ensure smooth operation at a higher rotational speed.

For maintaining a stable voltage at the terminal, an excitation system is employed by the synchronous machines, which may also include automatic voltage regulators (AVR). The introduction of a brushless excitation mechanism has reduced the necessity for a mechanical brush mechanism, which in turn helps with system reliability. Further, the auxiliary components, such as shaft and bearings, cooling systems, and damper windings, aid the system's stable, efficient, and reliable operation. In general terms, the physical design of synchronous machines is interlinked with the operating requirements, ensuring vigorous performance across a range of applications.

3.5.2 Rotating Magnetic Fields

For synchronous machines, the stator winding is supplied with three-phase balanced currents with 120° isolation. The currents make three sinusoidally distributed magnetomotive forces (MMFs) with a time-phase variation that results in a single rotating magnetic field. The underlying theory is that for every winding phase that generates pulsating MMF, which happens to be also aligned with its magnetic axis, as well as the vector sum of MMFs yielded by each phase, a resultant field is produced with a constant magnitude and even angular velocity. The rotating field is needed for generating rotor motion and determining electromagnetic torque through interacting with the excitation field of the rotor. The angular velocity associated with the rotating magnetic field is known as synchronous speed, and it relies on the provided frequency and the number of poles. The development of the rotating field can be investigated by observing three-phase currents as sinusoidal functions, which relate spatially and temporally to generating a magnetic field that remains constant in amplitude, however, rotates with synchronous speed. The subsequent field couples with the rotor conductors and, at the instance of appropriate excitation, secures the rotor into synchronization, further ensuring the machine provides steady torque production. This phenomenon is responsible for synchronous operation in AC machines.

The balanced currents for the machine are represented as:

$$\begin{aligned} i_a &= I_m \sin(\omega t), \\ i_b &= I_m \sin\left(\omega t - \frac{2\pi}{3}\right), \\ i_c &= I_m \sin\left(\omega t + \frac{2\pi}{3}\right) \end{aligned} \tag{3.17}$$

The resultant rotating MMF magnitude is almost constant and can be denoted by:

$$F_{\max} \approx \frac{3}{2} F_m \tag{3.18}$$

where F_m represents the peak MMF of one phase.

3.5.3 Electromotive Force (EMF) in Synchronous Machines

The interaction of the rotating magnetic field of the excited rotor with the stator's three-phase armature windings generates an electromotive force (EMF). As defined in the past section, the stator windings are placed in slots along the inner margin of the core lamination, and the rotating field covers these coils at a synchronous speed. As per Faraday's law, a time-fluctuating flux linkage for each phase of the winding creates a sinusoidal EMF, and the frequency of the EMF establishes the mechanical speed of the rotor and the number of poles. The allocation of the stator windings, pitch of the coils, and the harmonics in the rotating magnetic field affect the magnitude of the induced voltage. For a cleverly planned synchronous machine with allocated windings, the induced EMF approximates a pure sinusoid, which further enhances power quality and lowers harmonic losses. Further, the magnitude of line voltage alters depending on the air-gap flux per pole, the number of effective series turns, and rotors electrical speed, all together stressing the physical and electromagnetic relationships. Since the rotor field is regulated by DC excitation, modifying the excitation current influences the terminal voltage, which grants synchronous generators the ability to adjust system voltage and reactive power. The synchronized operation of the rotating magnetic field with the stator winding design forms the basis of EMF generation, which also links the machine's mechanical input to its electrical output.

The RMS value of the produced EMF per phase can be denoted as:

$$E_{\text{ph}} = 4.44 f\, \Phi\, T_{\text{ph}}\, k_w \tag{3.19}$$

where the winding factor is the consequence of distribution and pitch factors:

$$k_w = k_d\, k_p \tag{3.20}$$

The electrical frequency is associated with rotor speed as:

$$f = \frac{P n_m}{120} \tag{3.21}$$

And the flux linkage results in Faraday's law:

$$e(t) = -\frac{d\lambda(t)}{dt} \tag{3.22}$$

3.5.4 Three-Phase Equivalent Circuit

The steady-state performance for a synchronous machine operating under balanced three-phase conditions can be characterized by means of a per-phase equivalent circuit, which can record electromagnetic relations between the induced EMF, stator impedance, and terminal voltage. For this model, the internally produced EMF emerges after the synchronous reactance and stator resistance, giving a voltage relationship. This relationship between terminal voltage, internal EMF, and current is expressed as, based on Fig. 3.12:

$$\mathrm{V} = \mathrm{E_f} - jX_s\mathrm{I} - R_s\mathrm{I} \tag{3.23}$$

The synchronous reactance, which dominates the characterization, aids in establishing load angles, voltage regulation, and machine performance, relates both leakage flux impacts and the effect of armature reaction. This is represented by:

$$X_s = X_l + X_a \tag{3.24}$$

The power delivered per phase can be established from the equivalent circuit by analyzing the relationship between the terminal voltage and the link of the stator current. This relationship is essential for investigating machine loading and system integration, and is given by:

$$S_\phi = \underline{V}\,\underline{I}^* \tag{3.25}$$

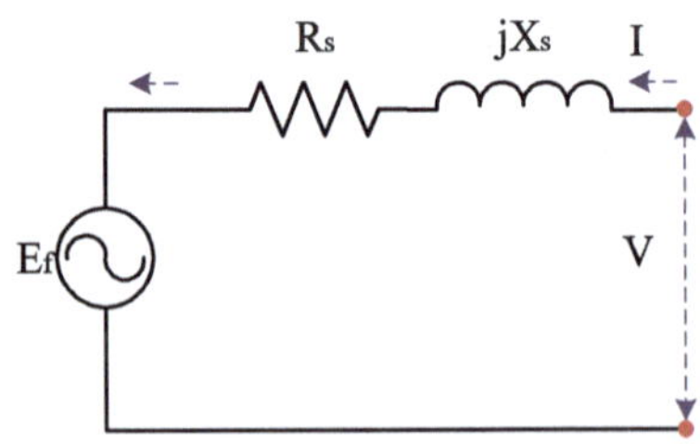

Fig. 3.12 Three-phase equivalent circuit for a synchronous machine

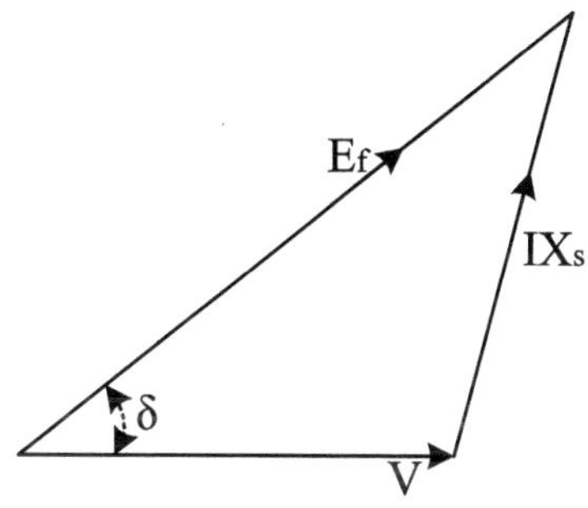

Fig. 3.13 Power angle for the synchronous machine

For analysis, interpretation of the magnitude relationship between the internal EMF and the terminal voltage is essential, specifically when estimating excitation levels or voltage regulation. This magnitude forms the voltage equation, which includes the load angle δ and is stated as:

$$= \sqrt{V^2 + (IX_s)^2 + 2VIX_s\sin\delta} \tag{3.26}$$

E_f This per-phase equivalent circuit therefore delivers a method to associate the physical structure of synchronous machines with their electrical performance, and provides evaluation of steady-state operation and reactive power control. The power angles for a synchronous machine are shown in Fig. 3.13.

3.5.5 Generator Active and Reactive Power

In a synchronous generator, the relations between the internal EMF and the terminal voltage influence the emergence of both active and reactive power to the electrical network. These power mechanisms are controlled by the magnitude of excitation, the load angle, and the machine's synchronous reactance. Active power is principally affected by the mechanical input and the torque angle between the internal EMF and the terminal voltage, while reactive power is primarily affected by the degree of field excitation. By altering the rotor excitation, the generator can supply or absorb reactive power, thus initiating voltage regulation and system stability. The fundamental illustrations for active and reactive power result precisely from the per-phase equivalent circuit and demonstrate the impact of electromagnetic into real and reactive power interchange with the grid.

The per-phase active power delivered by a synchronous generator is given by:

$$P = \frac{VE_f}{X_s}\sin\delta \tag{3.27}$$

This equation shows that active power depends primarily on the torque angle δ, while the magnitudes of the terminal voltage and internal EMF scale the power capability. Since δ considers mechanical–electrical coupling, rising mechanical input torque improves the load angle and therefore raises the active power output.

The reactive power exchanged by the generator depends on the relative magnitude of the internal EMF and the terminal voltage. Over-excited generators supply reactive power, while under-excited generators absorb it. The per-phase reactive power is expressed as:

$$Q = \frac{V}{X_s}(E_f \cos\delta - V) \tag{3.28}$$

This relation highlights the role of excitation control in regulating reactive power flow. If $E_f \cos\delta > V$, The generator delivers positive reactive power to the grid; if the opposite is true, it behaves like a synchronous condenser absorbing VARs. Thus, excitation systems and automatic voltage regulators (AVR) give operators direct control over the reactive power contribution of the machine.

Together, these active and reactive power expressions provide the foundation for understanding generator operating limits, stability boundaries, and grid-support capabilities. They show how mechanical input and excitation settings translate into controllable power outputs, integrating the generator into the wider power system.

3.6 PowerFactory Modelling

3.6.1 Modelling of Transformers

In PowerFactory, the drawing tools have different transformers as shown in Fig. 3.14, including a 2-winding transformer, an auto transformer, a step voltage regulator, a 3-winding transformer, a 4-winding transformer, a multi-winding transformer, a booster transformer, and a single-phase booster transformer.

As an example, a 2-winding transformer is installed between two busbars with 10 kV at the grid and 6 kV at the load side. Therefore, as shown in Fig. 3.15, two busbars are installed, then a transformer is connected in the middle of them.

After double-clicking on the transformer, the operational window of the transformer can be accessed as shown in Fig. 3.16. In this window, the number of transformers in parallel can be defined, which in this example is one. By clicking on the arrow in front of the type and selecting a new type, the window of the transformer's type can be accessed, as shown in Fig. 3.17.

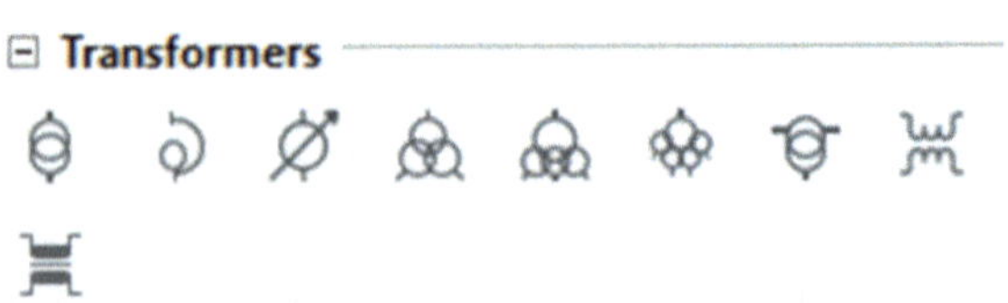

Fig. 3.14 Transformers in PowerFactory

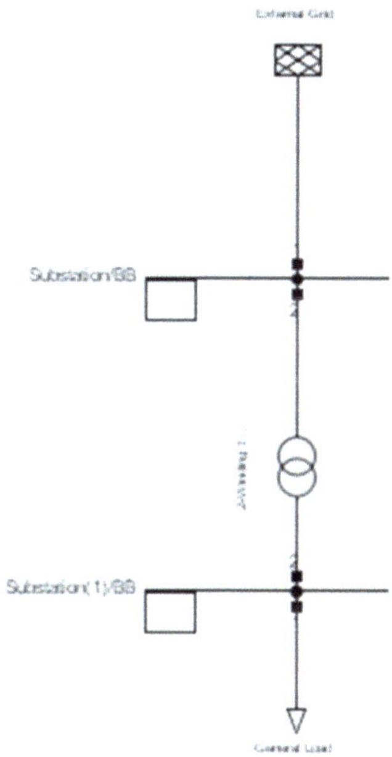

Fig. 3.15 A single-line diagram of a grid-connected load through a transformer in PowerFactory

Fig. 3.16 Operational window of the transformer

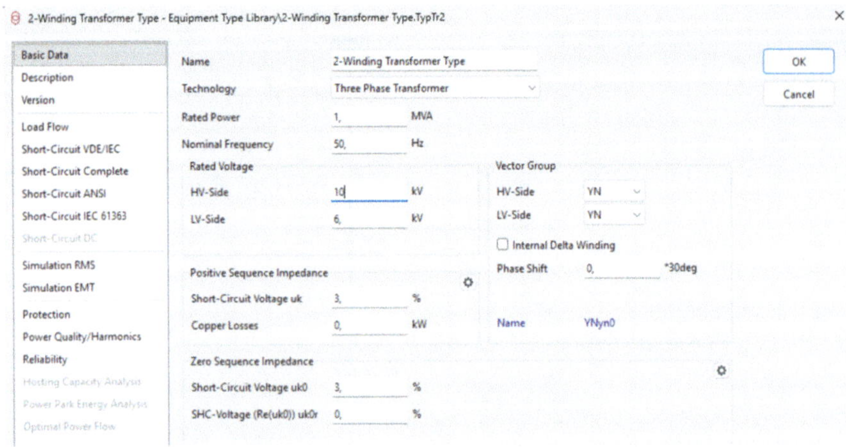

Fig. 3.17 Type window transformers

In the Basic data window, the ratings of transformers can be defined. The rated power can be set as 1 MVA, which is normally selected based on the load value, here is 0.6 MW. The HV side voltage is 10 kV, and the LV side is 6 kV.

For defining the tap changer in transformers, by clicking on the load flow tap, as shown in Fig. 3.18, it can be defined. For this example, each tap makes 2% voltage variations, including 7 taps, with 0 being the nominal one. So, additional voltage per tap should be 2%, minimum position −3, and maximum position +3.

After running the load flow, the results show a perfect operation of the example, as shown in Fig. 3.19. However, in practice, typically, the voltage from the grid can be changed due to the grid operation events. Therefore, the grid's voltage can be changed to 0.95 as shown in Fig. 3.20.

After running the load flow again, the voltage of the secondary of the transformer at the load side is also reduced to 0.95 p.u, which is on the border of standards, as shown in Fig. 3.21.

To improve the voltage level of the load side, by double clicking on the transformer and going to the load flow tap, the Tap position can be modified as shown in Fig. 3.22. For moving back the voltage level to 1 p.u, as each tap will change 2%, the tap position can be changed to −2.

Consequently, after running the load flows again, the voltage of the load is returned to 0.99 p.u, as shown in Fig. 3.23. However, the voltage of the primary side of the transformer is still 0.95 p,u due to the grid's operation voltage value.

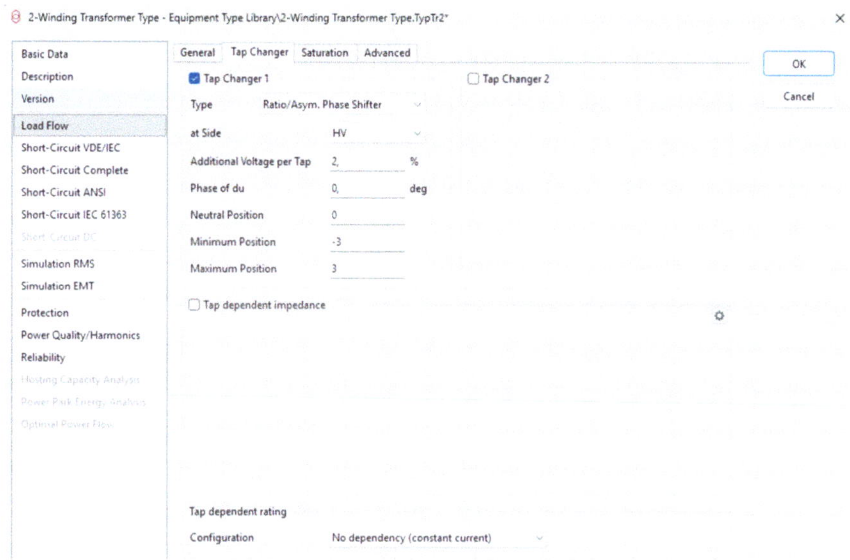

Fig. 3.18 Tap changers in transformers

Fig. 3.19 Load flow for transformer example

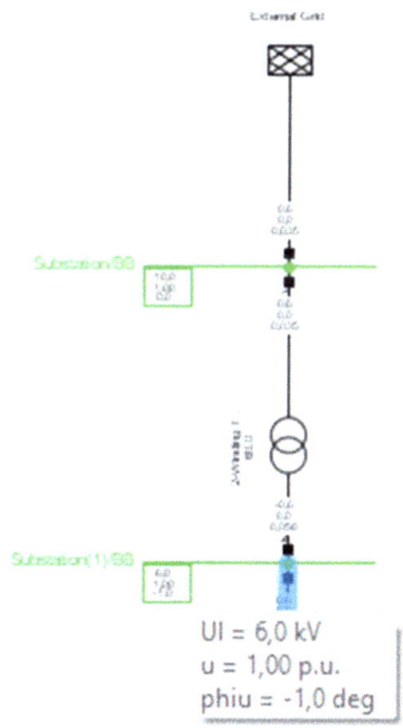

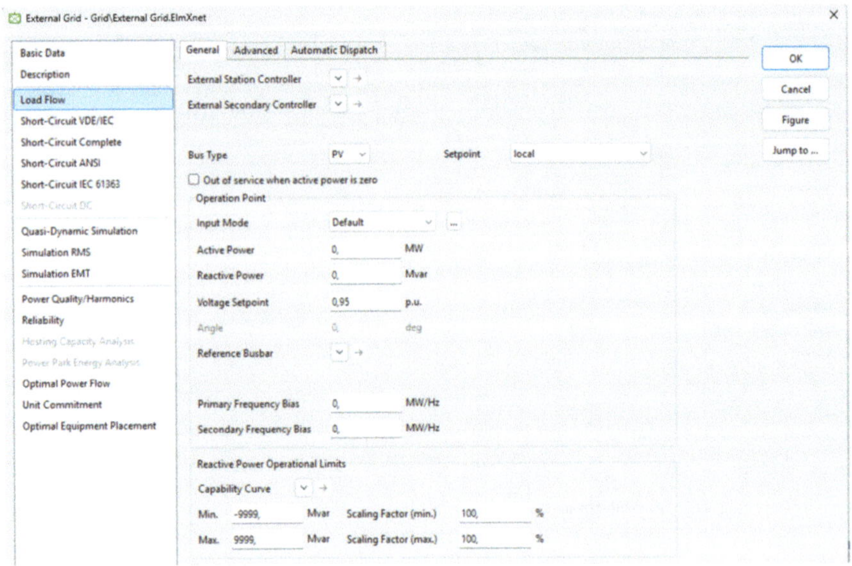

Fig. 3.20 Changing the grid's voltage set point

Fig. 3.21 Voltage drop in transformer example

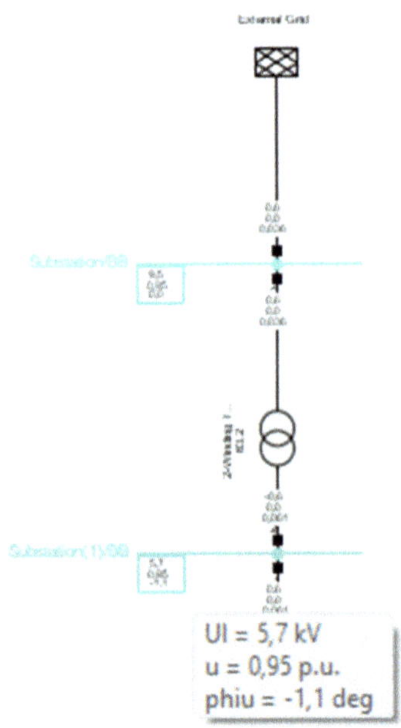

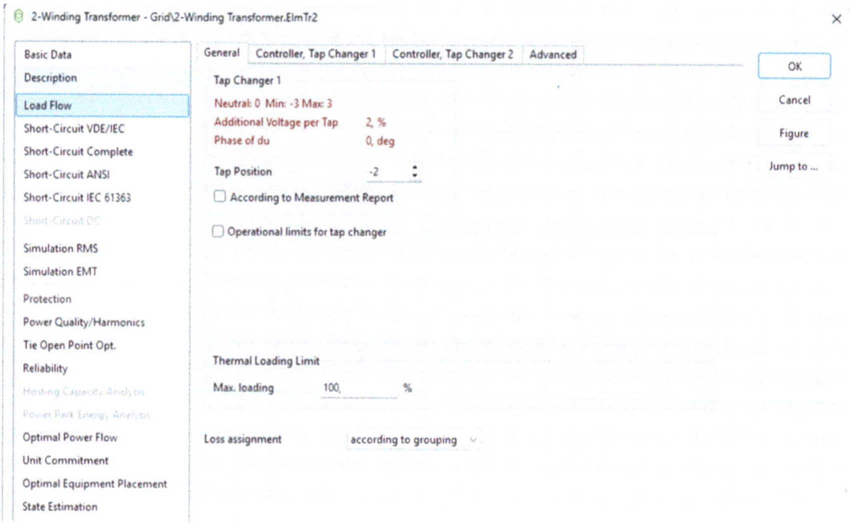

Fig. 3.22 Changing tap in transformer example

Fig. 3.23 Load flow of adjusted voltage in transformer example

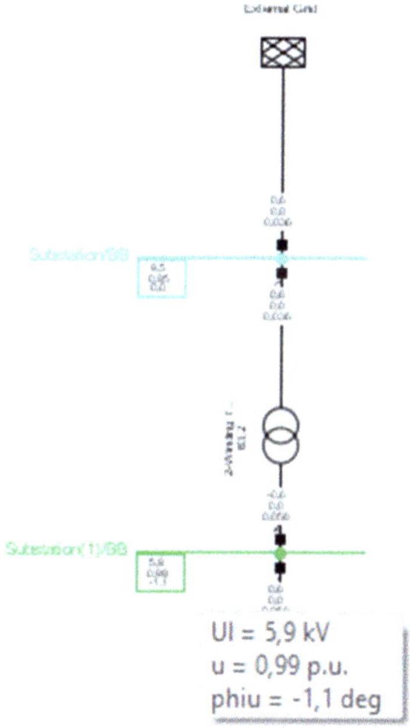

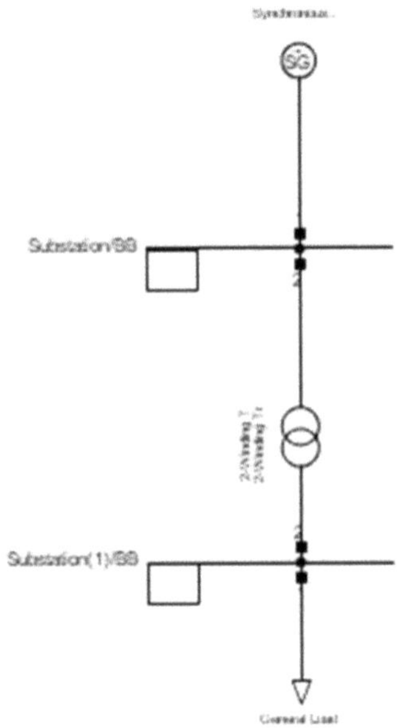

Fig. 3.24 Synchronous generator installation

3.6.2 Synchronous Machines Modelling

Synchronous generators in PowerFactory can be found in the drawing tools. In continuation of the previous example, a synchronous generator can be replaced with a grid, as shown in Fig. 3.24.

After installation of the synchronous generator, by double clicking on it, and making a new type, the window of Fig. 3.25 can be obtained. As the transformer is 1 MVA, the generator is also 1 MVA with a 10 kV rating voltage.

Then, after running the load flow, with the tap changer at position 0, the rating of current and power can be obtained as shown in Fig. 3.26. To observe the transients in PowerFactory, the RMS/EMT simulation can be performed. To do this, first the simulation results that are wanted to be observed should be defined by right-clicking on the generator and selecting result variables as shown in Fig. 3.27.

In the new window, the Phase current should be selected as shown in Fig. 3.28, and clicking ok. Then, by inserting a plot and adding the defined variable, the phase current of the generator can be observed as shown in Fig. 3.29.

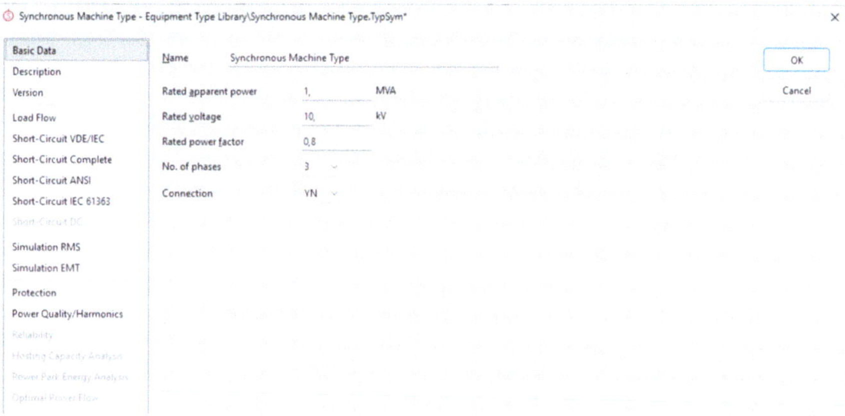

Fig. 3.25 Synchronous generator type window

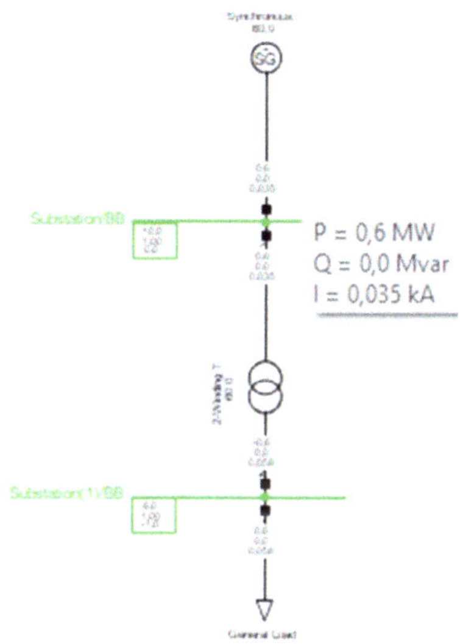

Fig. 3.26 Load flow of synchronous generator example

3.7 Summary

This chapter has presented the fundamental theory, modeling, and practical aspects of two cornerstone components of power systems: transformers and synchronous machines. It began with the critical role of transformers in stepping up or stepping down voltage levels to enable efficient long-distance transmission, reduce I2R losses, provide electrical isolation, supply neutral points in polyphase systems, and support accurate measurement through current and potential transformers. The

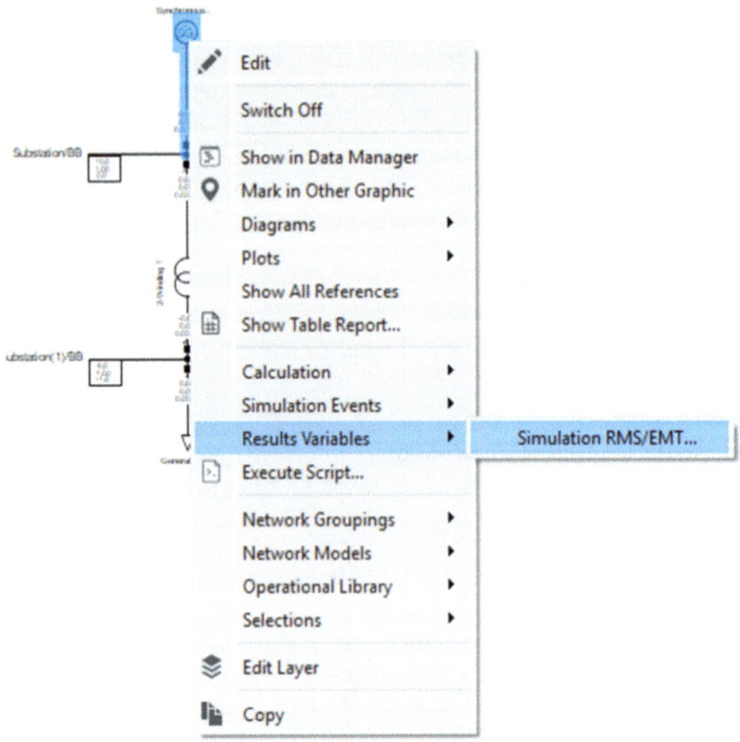

Fig. 3.27 Adding variables from a synchronous generator for the RMS/EMT simulation

ideal transformer model was introduced, covering voltage and current transformation ratios, power invariance, and impedance reflection. This was followed by the development of a realistic equivalent circuit that accounts for winding resistances, leakage reactances, core losses, and the magnetizing branch, enabling precise analysis of efficiency, voltage regulation, and performance under load. The chapter then extended to the working principles and detailed modeling of synchronous machines, from basic construction and operating principles to accurate equivalent-circuit representations suitable for steady-state and dynamic power-system studies. Finally, step-by-step modeling tutorials in PowerFactory—DIgSILENT were provided for both transformers and synchronous machines to reinforce theoretical concepts with practical simulation skills. The knowledge gained in this chapter forms an essential foundation for subsequent chapters on power flow, short-circuit analysis, protection, and stability studies.

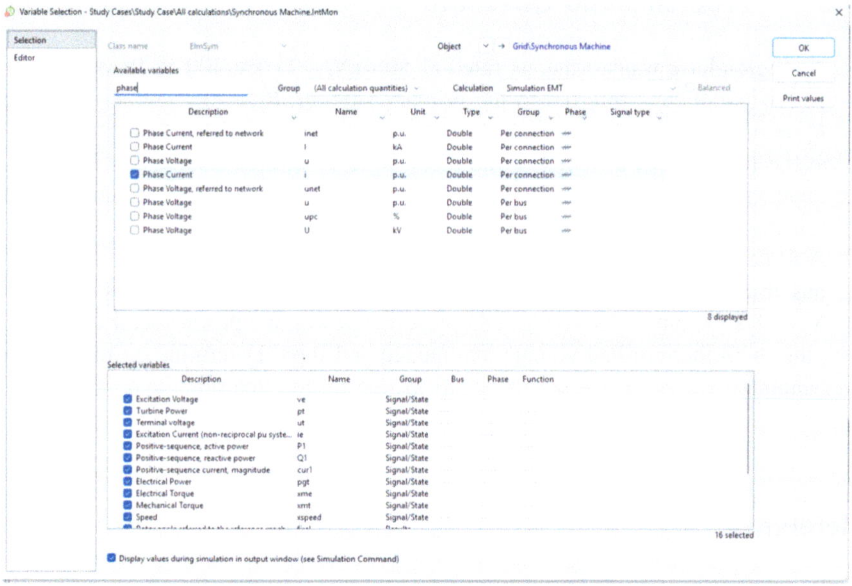

Fig. 3.28 Synchronous generator variables in RMS/EMT simulation

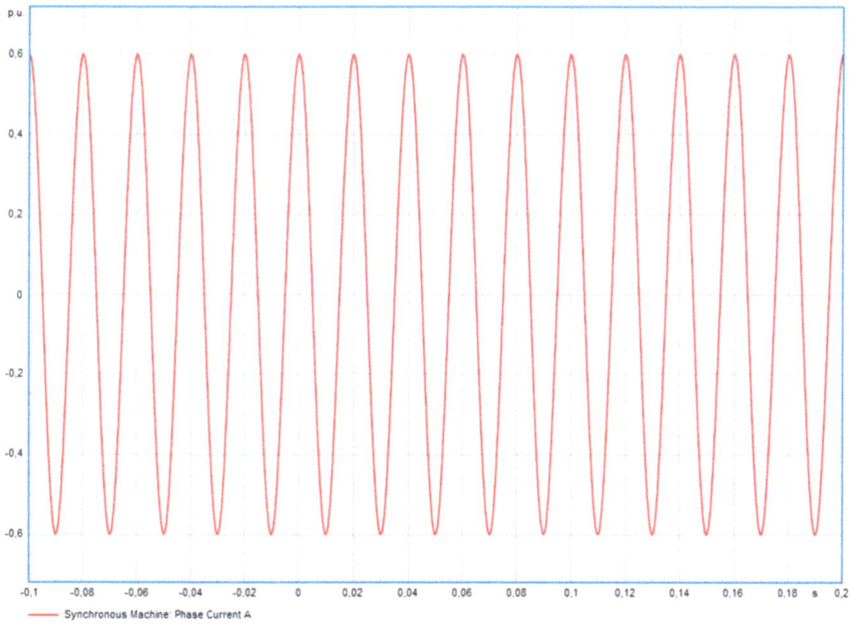

Fig. 3.29 Synchronous generator transient operation from EMT simulation

3.8 Problems and Questions

1. A single-phase transformer is rated at 500 kVA, 11 kV/400 V. Its equivalent-circuit parameters referred to the primary side are $R_1 = 0.25\ \Omega$, $X_1 = 1.2\ \Omega$, $R_c = 4500\ \Omega$, and $X_m = 3200\ \Omega$. The transformer supplies a load of 400 kVA at 0.8 lagging power factor while the secondary terminal voltage is held at its rated value. Calculate (a) the voltage regulation and (b) the efficiency of the transformer at this load condition.
2. A three-phase synchronous generator rated 50 MVA is connected to an infinite bus through a transmission line. The generator has a synchronous reactance $X_s = 1.2$ pu and is delivering 0.85 pu active power at a power factor of 0.9 lagging, with the terminal voltage maintained at 1.0 pu. Determine (a) the internal generated voltage E (in per unit) and (b) the torque (load) angle δ.

Reference

1. Power transformers. Energy Efficient Products. https://energy-efficient-products.ec.europa.eu/product-list/power-transformers_en

Transmission Systems

4

Abstract

This chapter has provided a comprehensive treatment of overhead transmission lines and underground cables, the essential links that transport electrical energy across power systems. It began by detailing the physical parameters of lines—series resistance, inductance, and shunt capacitance—including their calculation for single-circuit, bundled-conductor, and transposed configurations. Transmission lines were classified as short, medium, and long, with corresponding lumped and distributed equivalent-circuit models (nominal-π, nominal-T, and hyperbolic ABCD-parameter representations) derived for accurate performance evaluation. Key operational phenomena such as voltage drop, power losses, surge-impedance loading, Ferranti effect, voltage regulation, corona discharge, and skin effect were examined together with their impacts on efficiency, stability, and power quality. The distinctive features of cables (higher capacitance, lower inductance, and special insulation requirements) were contrasted with overhead lines, highlighting modeling challenges for underground and submarine applications. Finally, detailed step-by-step modeling tutorials in PowerFactory—DIgSILENT were presented, enabling readers to build, parameterize, and analyze realistic line and cable models for load-flow, short-circuit, and dynamic studies. The concepts and practical skills developed in this chapter form a vital foundation for the power-flow, short-circuit, protection, and stability analyses covered in the following chapters.

Keywords

Transmission line • Cables • Inductance of cables • Ferranti effect

N. Bayati and M. A. Khan, *Power System Fundamentals with PowerFactory*,
https://doi.org/10.1007/978-3-032-32450-4_4

4.1 Introduction

Power transmission for an electrical system is essential for the operation of a modern power system, as it is crucial for bulk power transfer from the generation unit to the substations and further on to the loads. Generally, the power generation unit is located in an isolated location, and the consumption units are distant. To ensure minimum power loss, it is essential to perform the transmission through a high-voltage transmission line. Usual transmission voltages scale from 110 to 765 kV in case of an overhead lines, while some regions also commission even higher voltages for ultra-high voltage (UHV) systems.

Transmission systems must meet certain technical standards for achieving reliable and present an efficient power transfer capability while delivering satisfactory power quality and maintaining stability under power disruptions. The electrical behaviours of a transmission line are managed by parameters such as conductance (G), resistance (R), inductance (L), and capacitance (C), as illustrated in Fig. 4.1. These parameters affect power loss, voltage profile, distribution of current, transient nature, as well as electromagnetic interactions for each of the lines. Due to the affirmation phenomena, it is essential to evaluate the transmission line with both steady-state and dynamic analysis.

As power networks continue to develop, the function of transmission systems has become more substantial. Rising incorporation of renewable energy resources, such as solar, introduces further irregularity and uncertainty into the grid. Such energy source sites are usually in remote regions, such as offshore wind farms or desert solar fields, which can further increase dependence on long-distance transmission corridors. Modern grid developers should address operational challenges along with traditional parametric concerns. There are also issues related to intermittency, bidirectional flow of power, and enhanced flexibility requirements that need to be addressed.

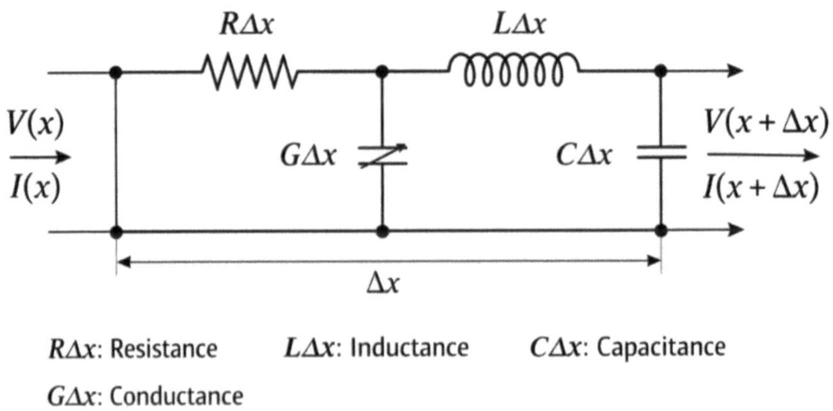

Fig. 4.1 Transmission line equivalent circuit

These challenges are always leading to innovations in transmission technology. For instance, High-voltage direct current (HVDC) systems provide outstanding controllability while reducing line losses over long distances, and enhanced integration qualifications for asynchronous grids. Also, smart grid technologies, such as advanced protection schemes, phasor measurement units (PMUs), and real-time monitoring systems, also permit operators to improve situational awareness, avoid cascading failures, and improve power transfer for power networks.

Recognizing the electromagnetic and electrical characteristics of transmission lines is essential to all these improvements. Precise modelling of line parameters, which is established through analytical techniques and based on computational tools, tends to present functionality for predicting system behaviours under normal and disturbed conditions. Such evaluations impact choices related to system planning, protection design, and stability assessments. The analysis also provides justification for phenomena like overvoltage, resonance, corona discharge, and electromagnetic interference.

Eventually, the analysis of transmission systems bridges fundamental electrical concepts with practical engineering reflections, playing a pivotal role in safeguarding modern power grids and ensuring their efficiency, reliability, and resilience.

4.2 Parameters of Transmission Lines

As discussed in the section before, transmission lines can be explained by four distributed parameters, which are resistance, inductance, capacitance, and shunt conductance. Each of these parameters is shaped by conductor material, geometry, spacing, and the medium surrounding it. They mutually influence performance aspects such as voltage, reactive power, power losses, and electromagnetic field. Resistance represents the real power dissipation, whereas inductance governs magnetic coupling and reactive effects. Further, the capacitance influences charging currents and electric field distribution, whereas conductance demonstrates leakage through insulation.

These parameters, when operating together, form transmission line modelling which can determine line behaves under steady-state and transient conditions. Such interaction impacts propagation characteristics, switching responses, and system stability. Hence, precise depiction of these parameters is critical for designing lines that accomplish reliable, efficient, and economical power transfer while ensuring the effective demands of modern power systems.

4.2.1 Resistance of Conductors

The resistance of a transmission line signifies the opposition to the flow of electric current, and it is responsible for real power losses, which are demonstrated as heat dissipation. These losses not only reduce efficiency but also affect thermal limits, mechanical sag, and conductor aging in the long term. For a low frequency or under DC operation, resistance can be approximated using the standard expression, which can be represented as:

$$R_{\mathrm{dc}} = \rho \frac{l}{A} \tag{4.1}$$

where material resistivity is denoted by ρ, conductor length is represented as l, and A is the cross-sectional area. The current distribution is assumed to be uniform, while assuming the effect of electromagnetic fields to be negligible. The selection of conductor material, such as aluminum, influences electrical conductivity, cost, weight, and resistance against corrosion.

When operating in AC environment, the operational resistance increases because of the non-uniform current distribution within the conductor. The skin effect will also force current to transfer toward the conductor's surface when there is an increase in frequency. The phenomenon happens because of the varying magnetic field, which happens to induce eddy currents within the conductor that tend to oppose the current flow within the inner layer and substantially reduce the effective area of conduction. The penetration depth of the current is represented by skin depth (δ), which is given by:

$$\delta = \sqrt{\frac{2\rho}{\omega\mu}} \tag{4.2}$$

where the angular frequency is represented by ω and magnetic permeability is denoted by μ. Even though the impact of skin effect is mild for an overhead conductor, it can have a significant impact for a large diameter conductor with high-frequency transients, and bundled conductor configurations.

The proximity effect additionally reduces current distribution when multiple conductors are close to each other, as in three-phase systems. The magnetic fields generated by adjoining conductors interfere with the current in each phase, driving current into confined regions distant from adjoining phases. This effect raises AC resistance and is significant for high-current lines, reduced tower geometries, and extra-high-voltage transmission systems, where bundles may hold up to eight sub conductors. Analytical modelling for the proximity effect needs resolving Maxwell's equations for a conductor and is frequently considered using Carson's equations, geometric mean radius (GMR), or finite element analysis (FEA) for accuracy. Temperature is also a substantial factor affecting conductor resistance. As the temperature of metal conductors increases, atomic vibrations intensify and delay electron mobility, which results in higher resistivity. The temperature dependence of resistance is normally represented by:

$$R_{\mathrm{ac}}(T) = R_{\mathrm{ac}}(T_0)[1 + \alpha(T - T_0)] \tag{4.3}$$

where the temperature coefficient is α and the temperature reference is T_0. As the overhead conductors reach a temperature of 40 °C under light loading and can go up to 100 °C during peak loading, precise thermal modeling is crucial for defining the permissible ampacity, for preventing excessive sag, and maintaining safe clearances to ground. Dynamic line rating (DLR) systems depend on real-time temperature measurements and environmental data to adapt line limits established on thermal behavior.

Selecting the suitable conductor type with consideration for electrical and mechanical performance is critical for a practical application. Stranded conductors, such as Aluminum Conductor Steel Reinforced (ACSR), are used significantly for high-voltage transmission because of their superior mechanical strength and good conductivity. The steel core grants high tensile strength, ensuring long lifespans and endurance to ice and wind loading on the cable, whereas the surrounding aluminum strands offer excellent electrical performance while also reducing the weight. Variants such as All-Aluminum Alloy Conductor, Aluminum Conductor Steel Supported, and Aluminum Conductor Composite Core are selected based on specific requirements such as high-temperature performance, reduced sag, or corrosion resistance.

The modelling of conductor resistance is vital for power flow studies, calculation of losses, and transient analysis. In addition, resistance significantly affects fault current calculations, influencing the selection of protective relays and circuit breakers. During fast transient events such as switching surges, frequency-dependent resistance becomes important due to enhanced skin effect, requiring more advanced modeling techniques.

Ahead of electrical concerns, conductor resistance causes thermal stress, material aging, and long-term reliability. Higher temperatures accelerate creep in aluminum conductors, increase mechanical sag, and can degrade composite cores. As a result, thermal–mechanical modeling along with electrical resistance analysis is necessary to safeguard operationality under varying environmental conditions.

In summary, conductor resistance in transmission lines is a multifaceted parameter influenced by material properties, geometric configuration, operating frequency, and temperature. An absolute understanding of these effects, along with advanced modeling methods, is necessary for designing efficient, reliable, and cost-effective transmission systems. Resistance not only impacts electrical performance but also dominates thermal behaviour and mechanical integrity.

4.2.2 Inductance of Transmission Lines

Inductance in overhead transmission lines originates from the magnetic flux which happens to link a conductor while alternating current flows through it. As current alternates, it creates a time-dependent magnetic field surrounding the conductor, and the varying flux causes a voltage that opposes the alternating current in agreement with Faraday's and Lenz's laws. This effect is present for all AC conductors and is measured as inductance per unit length. In a transmission line, the total inductance is a resultant of both self-inductance and mutual inductance.

Because three-phase lines consist of multiple closely spaced conductors, the impact of mutual inductance is significant and must be considered for line design and analysis.

For a single conductor situated in free space, the inductance per unit length can be expressed as:

$$L = 2 \times 10^{-7} \ln\left(\frac{D}{r'}\right) \tag{4.4}$$

where the distance between the conductor and its return path is denoted by D, and r′ is the GMR. The GMR signifies the equivalent radius which relations for internal flux linkages within the conductor. Since the current distribution for a realistic conductor is not uniform, the GMR is smaller compared to the physical radius. In the case of a solid round conductor, the effective radius is:

$$r' = 0.7788\,r \tag{4.5}$$

where the physical radius is denoted by r. This relation incorporates the logarithmic averaging, which is needed to precisely represent the internal magnetic field. For a three-phase line, each conductor conducts a phase current, and its magnetic field cooperates with the other two phases. As a result, the inductance of a given phase not only depends on the geometry of the conductor but also on the spacing between the three conductors. To consider the cumulative effect of three mutual inductances, the inductance for each of the phase is stated in terms of an equivalent spacing, and can be evaluated as the geometric mean of separation for the actual phase. Even though a balanced three-phase configuration streamlines the analysis, transmission lines rarely have absolute equal spacing due to structural and environmental limitations. To mitigate inductance imbalance and maintain an equal impedance in all phases, transmission lines are periodically transposed along the length so that each phase captures an equidistant physical position. It minimizes unbalanced voltages, reduces circulating currents, and improves system stability.

Conductor gaps have a strong influence on inductance. When conductors are closer to each other, the mutual flux linkage increases, which in turn increases the overall inductive reactance. Higher reactance causes a larger reactive voltage drop, increases absorption for reactive power, and lowers efficiency. As the maximum power transfer capacity for a transmission line is inversely proportional to its reactance, a rise in inductance limits the amount of real power that can be transferred over long distances. Hence, because of the above reasons, transmission lines which are operating at high and extra-high voltages are usually designed with large spacing between the phases and mostly employ bundled conductors as the bundling increases the effective GMR for the phase, resulting in inductance and reactance reduction, as well as lower corona discharge effects, and also improves power transfer capability along with voltage regulation.

Further, the inductive properties of a transmission line also affect reactive power behavior. Since inductive reactance which triggers a voltage drop is directly proportional to ωLI, the length of the lines significant reactive voltage drops which must then be recompensed through shunt capacitors, series capacitors, or power electronics components such as STATCOM devices. Furthermore, inductance initiates the line to absorb reactive power, specifically in long EHV lines, impacting system operation under both light and heavy loading conditions. Decreasing inductance through conductor arrangement and bundling leads to voltage stability and angular stability.

Thus, inductance is a crucial parameter for transmission line design, affecting electrical performance, stability, power transfer competence, and system reliability. A thorough selection of conductor shape, spacing, bundling arrangement, and transposition approach is needed for safeguarding efficient operation.

4.2.3 Capacitance of Transmission Lines

Capacitance in transmission lines results from the electric field that occurs amongst conductors of different potentials and between conductors and the ground. Every time two conductors are separated by a dielectric medium, an electric field is established between them, giving rise to electrostatic energy storage. For overhead transmission lines, this dielectric is primarily air, whereas for the underground cables, the dielectric can be solid or fluid insulation with tends to have much higher permittivity. The capacitance magnitude depends upon the geometry of the conductors, their comparative spacing, and the electrical properties of the medium that surrounds them. In case of a single-phase overhead line containing two parallel conductors, the per unit length of the capacitance can be expressed as:

$$C = \frac{2\pi\varepsilon}{\ln\left(\frac{D}{r}\right)} \tag{4.6}$$

where the permittivity of air is denoted by ε, D represents the distance between the conductors, whereas the radius of the conductor is denoted by r. As the overhead conductors are set apart, and air has a relatively low permittivity, the subsequent capacitance is small and can often be negligible. However, for high-voltage and extra-high-voltage overhead lines, the increased conductor height and long distances make capacitance a significant parameter, which may also influence steady-state and transient performance.

For an underground, capacitance is one of the influential electrical characteristics. Cable conductors are enclosed within high-permittivity lining placed only a few centimetres from the opposing conductor, bringing in capacitance values that are larger in magnitude than the overhead lines. As a result, cable charging currents become considerable even for adequate operating voltages, affecting voltage regulation, reactive power behavior, and tolerated cable length without compensation. The high dielectric constant of common cable insulation materials, such as cross-linked polyethylene, increases the electric field displacement compared to air, thus increasing capacitance and raising its effects on system operation.

Capacitance also plays a critical role in the transient and surge behavior of transmission systems. Because capacitance stores energy in the electric field, it interacts with line inductance to determine the pace of travelling waves and the characteristic impedance of the line. The transmission of lightning surges, switching surges, and fault-initiated transients is administered by the distributed RLC parameters of the transmission line, with capacitance influencing both the attenuation and distortion of these waves. In high-voltage networks, accurate modeling of line capacitance is essential for insulation coordination, surge arrester selection, and transient stability studies.

Although capacitance is generally beneficial from a reactive power perspective, it also presents challenges in long, lightly loaded lines, underground cable systems, and interconnected grids, wherever reactive power balance must be maintained. In some cases, shunt reactors are installed at line terminations to absorb excess reactive power delivered by line capacitance. Such compensation benefits control the Ferranti effect, maintain acceptable voltage profiles, and prevent overvoltage during normal and emergency conditions.

Overall, capacitance is a fundamental and mandatory characteristic of transmission lines, affecting steady-state operation, voltage regulation, reactive power flow, and transient performance. Its impact varies widely amongst overhead lines and underground cables due to differences in geometry and dielectric properties, but in all cases, an appropriate understanding and management of line capacitance is essential for designing reliable, efficient, and stable electric power transmission systems.

4.3 Characteristics and Performance of Transmission Lines

Transmission lines exhibit different electrical behaviour depending on their physical length in relation to the wavelength of AC power. Based on standard engineering practice, transmission lines are categorized as short, medium, or long lines. Each category uses a specific equivalent circuit representation to capture electrical characteristics. The performance of these lines is depicted using voltage regulation, power transfer capability, reactive power balance, and characteristic impedance.

4.3.1 Transposition

In three-phase transmission systems, the physical arrangement of conductors is almost never perfectly symmetrical. Tower geometry, mechanical constraints, and required safety clearances largely dictate the vertical and horizontal spacing between conductors. Because electrical parameters such as inductance and capacitance depend directly on conductor spacing and position relative to ground, each phase of a line ends up with slightly different electrical characteristics. Even when supplied by a perfectly balanced source, these small parameter disputes cause unequal voltage drops, phase-voltage imbalance, and nonuniform power sharing.

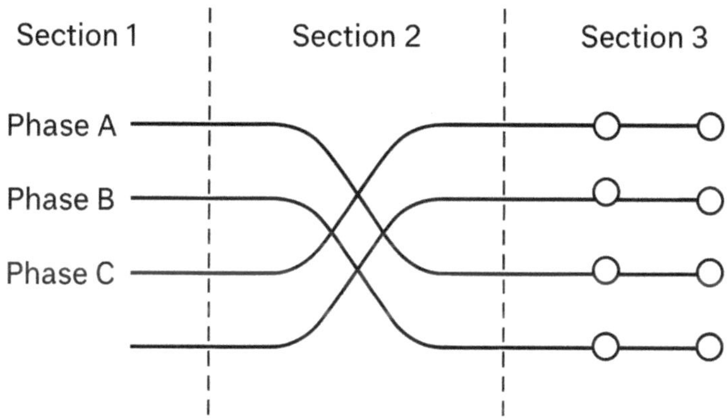

Fig. 4.2 A common transposition scheme

To mitigate these geometric irregularities, long transmission lines are transposed. Transposition is the intentional rotation of conductor positions along the line so that each phase occupies every physical position for an equal distance. Typically, the total line length is divided into three equal sections. At the end of each section, the conductors are rearranged so that Phases A, B, and C shift positions in a repeating cyclic pattern as shown in Fig. 4.2.

Through this spatial rotation, the average conductor spacing, phase-to-phase and phase-to-ground, becomes identical for all phases. As a result, each phase exhibits the same average inductance and capacitance, effectively restoring electrical symmetry even though no single point along the line is perfectly symmetrical.

This equalization significantly improves system performance. It enhances voltage balance, reduces neutral shift, and helps maintain consistent power sharing among phases. It also ensures that analytical tools such as steady-state power-flow studies, symmetrical component designs, and fault studies can reliably assume a balanced three-phase system. For long-distance and high-voltage lines, transposition is essential for maintaining stable, efficient, and predictable grid operation.

4.3.2 Bundling

GMR is a fundamental parameter used to characterize the electromagnetic behavior of transmission-line conductors. As alternating current does not distribute uniformly across the conductor cross-section, especially at higher power frequencies where internal flux linkages become substantial, the effective radius, which regulates inductance, is smaller than the physical radius. In practical terms, a larger

GMR reduces the self-inductive component of the line, lowering inductive reactance and improving the line's electrical performance. Typically, bundling can be between 2 and 6 conductors, as shown in Fig. 4.3.

Bundled conductors take advantage of this principle by combining two or more sub-conductors per phase. These sub-conductors are arranged symmetrically, spaced apart by rigid spacers, and electrically connected at both ends so they function as a single-phase conductor. The introduction of multiple sub-conductors creates a geometric form whose effective radius is significantly larger than that of a single conductor. The relationship between reactance and GMR is given by:

$$X_L \propto \ln\left(\frac{D}{GMR}\right) \tag{4.7}$$

This equation underlines the importance of rising GMR, as GMR grows, the denominator within the logarithmic term increases, causing the overall reactance to decrease. Lower reactance permits the transmission line to carry additional real power without excessive phase angle differences, reduces reactive power losses, and improves overall voltage stability. This becomes especially valuable in long-distance transmission corridors, where line impedance has a major impact on system performance.

In addition to lowering inductive reactance, bundled conductors also reduce the electric field intensity at the conductor surface. The increased effective radius spreads the electric field over a larger area, directly mitigating corona discharge. Corona produces audible noise, radio interference, and power loss, all of which are undesirable for system reliability and environmental impact. By increasing GMR and lowering surface electric stress, bundled conductors substantially reduce these effects.

Different transmission voltages and operational requirements lead to various bundling configurations. Two-bundle arrangements are commonly used for 220–345 kV lines, where corona effects are moderate, but reactance reduction is still beneficial. Three-bundle and four-bundle configurations are standard for 400–500 kV systems, offering a balance between electrical performance, mechanical

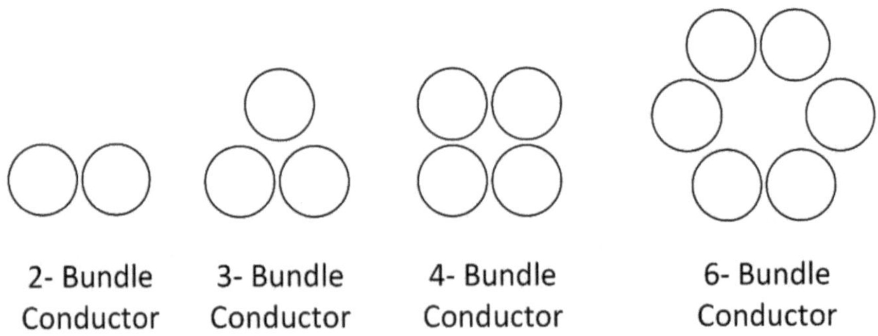

Fig. 4.3 Bundled conductor configuration

stability, and cost. At the highest voltage levels (765 kV AC), utilities often adopt six-bundle or even eight-bundle configurations. These larger bundles are necessary to control surface electric fields, manage conductor heating, reduce audible noise, and maintain acceptable levels of corona loss under extreme voltage stress. As the number of subs-conductors increases, the effective GMR rises steeply, producing large gains in both inductive performance and corona mitigation.

The choice of bundle type also depends on mechanical and environmental considerations. Larger bundles provide improved aerodynamic stability. Spacers maintain the geometry, preventing sub-conductors from clashing during storms or under wind-induced oscillations. However, additional sub-conductors and spacers increase installation complexity and cost, so network planners must evaluate trade-offs between performance, reliability, and budget.

Ultimately, GMR sits at the centre of both inductive behaviour and corona performance in transmission lines. Bundling stands as one of the most effective engineering strategies for enhancing line performance, and understanding its relationship with GMR is essential for designing high-voltage transmission networks.

4.3.3 Short Transmission Lines

Short transmission lines are typically defined as overhead lines less than about 80 km in length. At these distances, the line's shunt capacitance is extremely small because the conductors are spaced widely apart, and the electric field between each conductor and ground is weak. As a result, capacitance contributes negligibly to current flow and reactive power. The line can be modeled using a simple series impedance:

$$Z = R + j\omega L \tag{4.8}$$

Since there is no shunt capacitance component, the voltage and current relationship along the line is straightforward. The sending-end voltage is simply the receiving-end voltage plus the voltage drop across the line impedance:

$$V_s = V_r + IZ \tag{4.9}$$

It can also be represented as in Fig. 4.4. This simplified model is sufficient for MV and some HV overhead lines where capacitance does not meaningfully influence voltage regulation or reactive power flow.

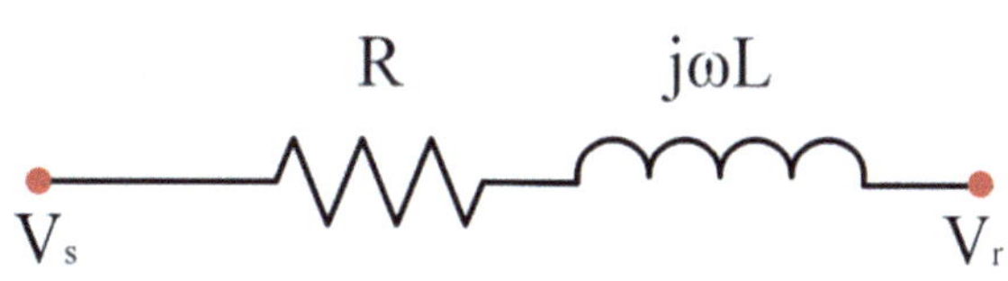

Fig. 4.4 Short transmission line model

4.3.4 Medium Transmission Lines

Medium lines, ranging from 80 to 250 km, exhibit measurable capacitance that must be included. The nominal π-model and nominal T-model are used for analysis. The π-model consists of a series impedance and two shunt capacitors at each end. The ABCD parameters for the nominal π-model are:

$$A = 1 + \frac{YZ}{2}, B = Z, C = Y\left(1 + \frac{YZ}{4}\right), D = A \tag{4.10}$$

where Z is the series impedance and Y is the shunt admittance. These models provide accurate voltage and current relations and are widely used in steady-state simulations.

4.3.5 Long Transmission Lines

Long lines, exceeding 250 km, require a distributed parameter model to account for the continuous nature of R, L, C, and G. The analysis uses hyperbolic functions. The propagation constant is defined as:

$$\gamma = \sqrt{(R + j\omega L)(G + j\omega C)} \tag{4.11}$$

and the characteristic impedance is:

$$Z_c = \sqrt{\frac{R + j\omega L}{G + j\omega C}} \tag{4.12}$$

The general voltage-current relations at the sender and receiver ends are:

$$\begin{aligned} V_s &= AV_r + BI_r \\ I_s &= CV_r + DI_r \end{aligned} \tag{4.13}$$

with:

$$\begin{aligned} A &= D = \cosh(\gamma l) \\ B &= Z_c \sinh(\gamma l) \\ C &= \frac{1}{Z_c} \sinh(\gamma l) \end{aligned} \tag{4.14}$$

These expressions describe wave propagation, reflections, attenuation, and phase shifts along the line.

4.3.6 Surge Impedance

The surge impedance, also known as the characteristic impedance of a transmission line, is a fundamental parameter that describes the natural relationship between voltage and current for a traveling electromagnetic wave along a line. For an ideal or lossless transmission line, the surge impedance depends only on the inductance L and capacitance C per unit length. It is defined as:

$$Z_c = \sqrt{\frac{L}{C}} \tag{4.15}$$

This impedance represents the value at which the reactive effects of inductance and capacitance are perfectly balanced. When a line is terminated with its surge impedance, the inductive and capacitive reactive powers cancel each other, resulting in no net reactive power exchange with the sending or receiving end. Under this condition, the line operates at its "natural" state, neither absorbing nor supplying reactive power.

A related concept is the Surge Impedance Loading (SIL), which represents the real power transferred when the line is loaded such that the receiving-end power factor is unity and the load impedance matches the surge impedance. SIL is a convenient metric used in system planning to assess the natural loading capability of HVAC transmission lines. It is calculated as:

$$SIL = \frac{V^2}{Z_c} \tag{4.16}$$

where V is the line's phase-to-neutral voltage for single-phase or line-to-line voltage for three-phase systems. At this loading level, the line is said to be "naturally loaded" and operates without requiring reactive power support from the system.

Because of these characteristics, SIL is frequently used in transmission planning to compare lines of different voltages and design configurations, evaluate their natural loading limits, and determine the need for shunt reactors, capacitors, or FACTS devices. Higher-voltage lines inherently have higher SIL values, allowing them to transmit greater power levels before encountering voltage stability limitations.

4.3.7 Ferranti Effect

The Ferranti effect is a well-known phenomenon in power transmission where the receiving-end voltage becomes higher than the sending-end voltage under light-load or open-circuit conditions. This counterintuitive voltage rise occurs primarily on long transmission lines—typically EHV overhead lines and, even more prominently, underground cables. The effect arises from the distributed nature of line inductance and capacitance, which interact to shape voltage magnitude along the conductor when the load current is small.

In an unloaded or lightly loaded line, the shunt capacitance draws a charging current that leads the voltage by approximately 90°. As this capacitive charging current flows through the series inductance of the line, it produces a voltage rise along the line length.

A simplified qualitative relationship for the Ferranti voltage rise is:

$$\Delta V \propto \omega^2 LCV \tag{4.17}$$

This expression shows that the effect increases with system frequency, line inductance, line capacitance, and operating voltage. More accurately, long lines must be analyzed using the transmission line differential equations or the long-line model, where the receiving-end voltage for an open-circuit line is:

$$V_r = V_s \cosh(\gamma l) \tag{4.18}$$

For a lossless line, $\gamma = j\beta$, giving:

$$|V_r| = |V_s| \cos(\beta l) \tag{4.19}$$

Since $\cos(\beta l) > 1$ for long, lightly loaded lines, the receiving-end magnitude becomes greater than the sending-end magnitude—capturing the mathematical basis of the Ferranti effect.

The effect is far more pronounced in underground cables, which have very high capacitance per unit length compared to overhead lines. Even relatively short cables can experience noticeable receiving-end overvoltages when lightly loaded or energized with no load. This characteristic necessitates the use of shunt reactors, reactive compensation, and careful voltage-control strategies on long EHV and UHV cable systems.

4.3.8 Corona Discharge

Corona discharge is a localized ionization phenomenon that occurs when the electric field intensity at the surface of a conductor exceeds the dielectric strength of air. Under such conditions, free electrons in the surrounding air are accelerated to energies high enough to ionize nearby molecules, creating a region of partially ionized gas around the conductor. This ionization forms a faint luminous glow, particularly visible in low-light conditions, and is accompanied by audible noise. In addition to visual and acoustic effects, corona also produces ozone and nitrogen oxides due to high-energy collisions in the ionized region, which contribute to localized chemical activity around the conductors.

The onset of corona can be described using Peek's empirical formula, which gives the disruptive critical voltage:

$$V_0 = m_0\, g_0\, r\left(1 + \frac{0.301}{\sqrt{r}}\right) \tag{4.20}$$

where, r is the conductor radius, g_0 is the dielectric strength of air, and m_0 is the surface irregularity factor.

This expression shows that the corona onset voltage depends strongly on the conductor's sector size and condition. A larger radius reduces the surface electric field strength, thereby increasing the corona onset voltage. Surface contamination, roughness, corrosion, and stranded-wire geometry increase the effective electric field at microscale points, reducing the onset threshold. Weather conditions also play a significant role: rain, fog, and snow increase local field intensification and cause corona to occur at lower voltages.

Corona discharge leads to undesirable corona power loss, caused by energy dissipating in the ionized air around the conductor. The power loss increases significantly with operating voltage and becomes especially important in EHV and UHV systems. Corona also radiates electromagnetic noise, creating radio and audible interference that must be considered in both line design and environmental impact assessments. Because these effects become more pronounced as voltage increases, mitigation becomes essential for high-voltage transmission networks.

4.4 Cables

Underground cables have become increasingly important in modern power systems, particularly in countries aiming to reduce the visual impact of overhead lines and improve reliability. Compared to overhead lines, cables require significantly more insulation, mechanical protection, and thermal management. Their construction reflects these needs through multiple layers of dielectric material, metallic protection, fillers, and sheathing. The following subsections present the construction, classification, electrical properties, stress behavior, grading techniques, and losses associated with underground cables.

4.4.1 Construction of Cables

The construction of underground cables is more complex than that of overhead conductors due to the need for electrical insulation and mechanical protection. As illustrated in Fig. 4.5, a typical underground cable consists of several concentric layers forming a robust insulated assembly. At the core of the cable lies the conductor, typically made of copper or aluminum, chosen for their high conductivity and mechanical flexibility. The structure of the conductor in Fig. 4.5, which highlights both single-core and multi-core arrangements used in various voltage applications. Surrounding the conductor is the insulation layer, which emphasizes the use of PVC, rubber compounds, or XLPE depending on voltage level and environmental requirements. XLPE insulation, in particular, offers high dielectric strength and excellent thermal performance.

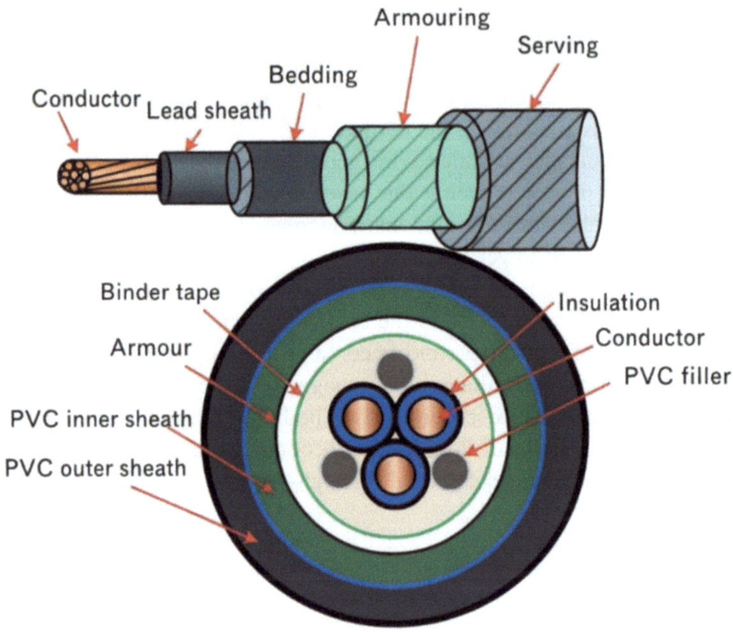

Fig. 4.5 Cross-section of an underground cable

In multi-core cables, empty spaces between insulated conductors are filled with PVC fillers, which help maintain the circular shape and structural stability of the cable. To hold the cores in place, a binder tape is applied, providing moisture resistance and preserving concentricity.

For mechanical protection, particularly during underground installation, cables are surrounded by steel wire armouring, which protects against crushing forces and external impacts. Between the armouring and metallic sheath lies a bedding layer, followed by the outer serving which safeguards the cable from environmental conditions and mechanical abrasion.

4.4.2 Classification of Cables

Cables can be classified according to several parameters, including their voltage rating, insulation material, number of cores, and intended application. Typical examples encountered in power systems include LSZH-insulated cables, XLPE-insulated cables, and low-voltage (LV) armoured distribution cables.

From a voltage perspective, Underground power cables are often grouped by their voltage rating, as this directly affects the design, insulation thickness, conductor construction, and mechanical protection requirements. Low-voltage (LV) cables operate up to 1 kV, while high-tension (HT) cables are rated up to 11 kV. The next category, super-tension (ST), covers the 22–33 kV range, followed by extra-high tension (EHT) cables used between 33 and 66 kV. Beyond this, extra-super tension (EST) cables are used for systems operating above 132 kV. These voltage groups serve as a fundamental basis for selecting appropriate cable designs across different parts of the power network.

The choice of insulation material has a major influence on a cable's dielectric behaviour, thermal performance, flexibility, and environmental suitability. Commonly used insulation types include PVC (polyvinyl chloride), which provides a cost-effective solution for many LV applications, and PE (polyethylene), which is known for high dielectric strength and excellent moisture resistance. XLPE (cross-linked polyethylene) is widely adopted in medium- and high-voltage networks due to its superior thermal stability and electrical strength, while EPR (ethylene-propylene rubber) is preferred when increased flexibility and heat resistance are required. In environments where fire safety is paramount, LSZH (low-smoke zero-halogen) insulation is used because it significantly reduces smoke and toxic emissions during combustion.

Cable classification also depends on the number of insulated conductors contained within a cable. Single-core cables contain one insulated conductor and are predominantly used in medium- and high-voltage systems. Their design minimises electromagnetic interference between phases, allows more flexible routing, reduces mechanical stresses from induced currents in metallic sheaths, and improves thermal performance under high load conditions. In contrast, multi-core cables contain two or more insulated conductors within a common outer sheath and are widely used in low-voltage systems as well as in control and instrumentation applications. Their configuration simplifies installation by combining multiple circuits into one cable, reduces space requirements, lowers installation costs, and allows efficient three-phase LV distribution with integrated earth conductors.

In addition to voltage level and insulation type, cables may also be categorised based on their intended application. Distribution cables are used throughout LV and MV networks to supply power to residential, commercial, and industrial customers. Submarine cables are specially designed for underwater transmission and require enhanced mechanical protection and moisture resistance. Cables intended for industrial environments often incorporate additional armouring and insulation to withstand mechanical stress, chemicals, or elevated temperatures. Instrumentation and control cables are designed for low-power signal transmission and are essential in automation, monitoring, and process-control systems. These application-based classifications help engineers choose cables that meet specific operational, environmental, and safety requirements. Summary of each cable type is presented in Table 4.1.

Table 4.1 Demonstrating different classification possibilities

Category type	Classification	Description/properties	Typical applications
Voltage level	Low voltage (LV)	Operates up to 1 kV	Domestic wiring, small commercial loads
	High tension (HT)	Operates up to 11 kV	Local distribution networks
	Super tension (ST)	Operates at 22–33 kV	Regional distribution, industrial feeders
	Extra high tension (EHT)	Operates at 33–66 kV	High-capacity distribution
	Extra super tension (EST)	Above 132 kV	Transmission systems
Insulation material	PVC	Economical, good mechanical strength	LV installations, buildings
	PE	High dielectric strength, moisture-resistant	Outdoor LV/MV systems
	XLPE	Superior thermal & dielectric performance	MV and HV power networks
	EPR	Flexible, heat-resistant	Industrial and specialised systems
	LSZH	Low smoke and low toxicity	Public buildings, tunnels, transport
Cable construction	Single-core	One conductor, low EM interference, good heat dissipation	MV/HV systems, long-distance feeders
	Multi-core	Multiple conductors under one sheath	LV distribution, control & instrumentation
Application type	Distribution cables	General LV/MV power distribution	Residential, commercial, industrial grids
	Submarine cables	Special underwater design	Offshore wind, island supply
	Industrial cables	High mechanical/ thermal resistance	Factories, process plants
	Control/ instrumentation cables	Carries low-power signals	Automation, monitoring, SCADA

4.4.3 Insulation Resistance

Insulation resistance is a fundamental parameter used to assess the health and electrical integrity of underground power cables. It represents the opposition offered by the insulating material to the flow of leakage current from the conductor to the surrounding sheath or environment. A high insulation resistance indicates that the dielectric material is effectively preventing current leakage, thereby preserving system efficiency and ensuring safe operation. Conversely, a reduction in insulation resistance is often associated with moisture ingress, thermal degradation, chemical contamination, or mechanical defects within the cable. Continuous monitoring and accurate determination of insulation resistance are therefore critical for preventing dielectric breakdown, reducing energy losses, and extending the operational life of cable systems.

The geometric configuration of the insulation layers surrounding a cable conductor is illustrated in Fig. 4.6. This figure also includes the differential cylindrical shell element used in the derivation of insulation resistance. To evaluate how the insulation resists leakage currents, the analysis considers an infinitesimal cylindrical layer of thickness dx located at a radial distance x from the centre of the conductor. The circumference of this layer, and therefore the effective area normal to the radial current flow, is expressed as:

$$A = 2\pi x l \tag{4.21}$$

For this differential layer, the elemental radial resistance dR_{ins} is defined based on the resistivity ρ of the insulation material:

$$dR_{ins} = \rho \frac{dx}{2\pi x l} \tag{4.22}$$

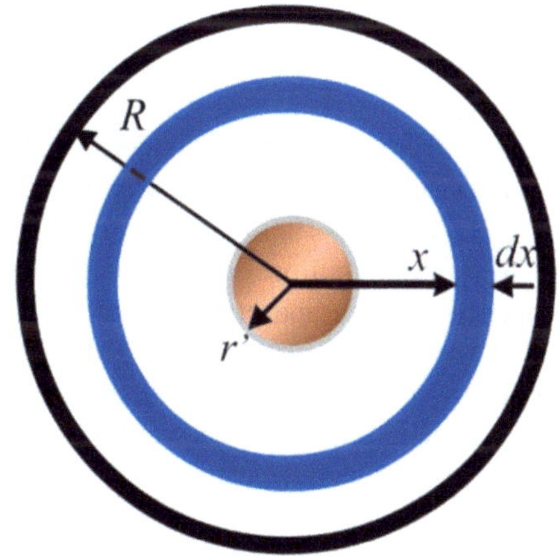

Fig. 4.6 Insulation resistance derivation diagram

To obtain the total insulation resistance of the cable, the elemental resistance is integrated between the conductor radius r and the inner radius of the outer sheath R:

$$R_{ins} = \int_r^R dR_{ins}$$
$$R_{ins} = \int_r^R \rho \frac{dx}{2\pi xl} \tag{4.23}$$

Solving the integral yields the classical logarithmic expression for insulation resistance:

$$R_{ins} = \frac{\rho}{2\pi l} \ln \frac{R}{r} \tag{4.24}$$

This equation shows that insulation resistance increases with both the resistivity of the dielectric material and the radial thickness of the insulation, while being inversely proportional to cable length. As such, thicker insulation layers, materials with high intrinsic resistivity, and shorter cable lengths all contribute to improved insulation performance.

Low insulation resistance is undesirable because it results in elevated leakage currents, dielectric heating, energy dissipation, and accelerated material aging. Prolonged exposure to these effects can ultimately lead to insulation breakdown and cable failure. For these reasons, careful selection of dielectric materials and appropriate insulation thicknesses is essential to ensure long-term reliability, operational safety, and compliance with industry standards for underground power cable installations.

4.4.4 Electric Stress and Economical Sizing

The radial electric field distribution within a cable's insulation is an essential parameter for determining dielectric performance and long-term reliability. In a single-core power cable, the electric field is not uniformly distributed across the insulation thickness; instead, it varies with radial distance from the conductor. Since electric stress is highest close to the conductor surface, a proper understanding of this distribution is crucial for selecting insulation materials with adequate dielectric strength and ensuring safe cable design. The electric stress distribution within the insulation. Figure 4.7 shows how the electric field is highest at the conductor surface and decreases radially.

For a cable with conductor radius r and inner sheath radius R, the electric field intensity at any radial position x within the insulation can be expressed as:

$$E = \frac{V}{x \ln \frac{R}{r}} \tag{4.25}$$

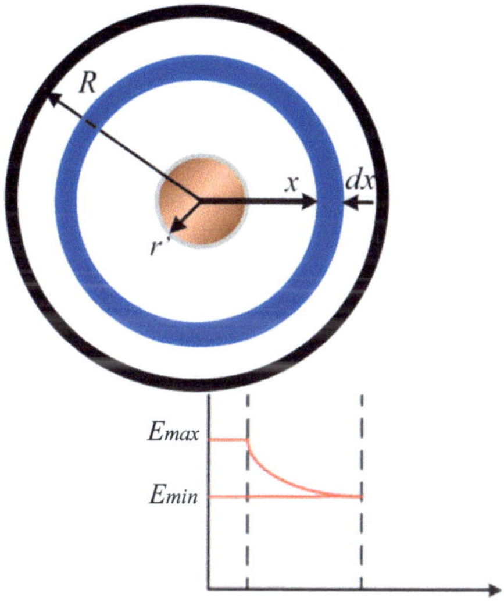

Fig. 4.7 Electric field distribution

where V is the operating voltage. This relationship shows that electric stress is inversely proportional to the radius x, meaning the stress decreases gradually as one moves away from the conductor towards the outer sheath.

Because of this inverse dependence, the stress distribution over the dielectric is inherently non-uniform. The maximum stress always occurs at the conductor surface, where the radius is smallest, while the minimum stress appears at the inner surface of the sheath. This behaviour is particularly significant in medium- and high-voltage cables, where marginal overstress near the conductor can accelerate insulation aging, initiate partial discharge, or lead to dielectric breakdown.

At the point where $x = r$, the maximum electric stress is obtained by substituting the conductor radius into the general equation:

$$E_{\max} = \frac{V}{r \ln \frac{R}{r}} \tag{4.26}$$

Similarly, at $x = R$, the minimum electric field within the insulation is:

$$E_{\min} = \frac{V}{R \ln \frac{R}{r}} \tag{4.27}$$

The ratio of maximum to minimum stress follows directly:

$$\frac{E_{\max}}{E_{\min}} = \frac{R}{r} \tag{4.28}$$

This ratio depends solely on the geometric proportion of sheath radius to conductor radius. A larger R/r ratio results in more severe non-uniformity, meaning the insulation near the conductor experiences disproportionately high stress. This effect is one reason why very thick insulation layers do not always translate into improved performance unless the geometric proportions are optimised.

Because the highest stress occurs directly at the conductor surface, the insulation material must possess a dielectric strength comfortably exceeding this maximum value. Failure to do so can cause local hot spots, intensify partial discharge activity, and lead to progressive deterioration of the insulation system. Therefore, managing electric stress through appropriate geometry, material selection, and stress-grading techniques is essential for ensuring long-term cable reliability.

As previously established, the electric field within a single-core cable is not uniformly distributed, and the maximum electric stress always occurs at the surface of the conductor. For safe and reliable operation, the dielectric strength of the insulation must exceed this maximum stress by an adequate safety margin. Because insulation materials become more expensive with increasing thickness, and because excessive stress near the conductor accelerates aging, it is beneficial to determine a conductor size that minimises the maximum electric stress for a given operating voltage and sheath radius. To determine the most economical conductor size, the goal is to identify the value of r that minimises $E_{\max}$. This is achieved by differentiating the denominator term $r \ln\left(\frac{R}{r}\right)$ with respect to r, setting the derivative equal to zero, and solving for the optimal radius:

$$\frac{d}{dr}\left(r \ln \frac{R}{r}\right) = 0 \tag{4.29}$$

Solving this condition yields the well-known result that the maximum electric stress is minimised when:

$$R = e\,r \tag{4.30}$$

This means that the sheath radius must be 2.718 times the conductor radius for the electric stress to be optimally distributed. The corresponding insulation thickness is then:

$$T_{\text{ins}} = R - r = 2.718r - r = 1.718r \tag{4.31}$$

This result provides a practical design guideline: for a given voltage rating and sheath dimension, the conductor size that yields the lowest possible maximum stress is one where the insulation thickness is approximately 1.718 times the conductor radius. This configuration ensures a balanced stress profile, reduces the risk of partial discharge at the conductor surface, and improves the long-term reliability of the insulation system.

In engineering practice, this relationship serves as a reference when optimising cable geometry for medium- and high-voltage applications. While actual conductor and sheath sizes are also influenced by thermal limits, manufacturing constraints, and standardised dimensions, the stress-optimised radius provides valuable insight into how geometric proportions affect dielectric performance.

4.4.5 Grading

Uniform distribution of electric stress within a cable's insulation is essential for achieving safe, efficient, and long-term operation. The dielectric material surrounding the conductor must be capable not only of withstanding the maximum electric stress but also of maintaining stability under varying electrical and thermal conditions. If the stress distribution is highly non-uniform—such as when the stress near the conductor is significantly higher than at the sheath—local regions of overstress can occur. These regions may initiate partial discharge, thermal degradation, or even catastrophic insulation breakdown.

To mitigate these risks, dielectric materials within cables are often selected and arranged to minimise the difference between maximum and minimum stress values. When the stress varies excessively across the insulation thickness, the required withstand capability increases, forcing the use of thicker insulation layers. This not only enlarges the cable diameter but also increases material cost and reduces thermal performance. Therefore, engineering techniques that help homogenise the electric field within the insulation are vital in medium- and high-voltage cable design.

The process used to achieve a more uniform electric-field distribution across the dielectric is known as grading. Grading techniques deliberately modify the permittivity or potential distribution within the insulation to flatten the electric-field profile and reduce peak stresses. One widely used method is capacitance grading, where two or more dielectric layers with different permittivities are placed concentrically around the conductor. Each layer alters the radial stress according to its permittivity, allowing engineers to shift higher stress to materials better suited to withstand it.

Consider a cable in which the conductor is surrounded by three dielectric layers with permittivities ε_1, ε_2, and ε_3, extending to radii r_1, r_2, and R, respectively. For a charge density q, the electric stress at any radial position x within a dielectric layer of permittivity ε_x can be expressed as:

$$E_x = \frac{q}{2\pi\varepsilon_o\varepsilon_x x} \tag{4.32}$$

This expression highlights the dual dependence of electric stress on radial distance and material permittivity. By assigning higher-permittivity materials closer to the conductor, the electric field can be reduced in the regions where stress is naturally highest. Conversely, lower-permittivity materials are placed toward the sheath, where stress levels are inherently lower.

Figure 4.8 illustrates the configuration of capacitive grading using concentric dielectric layers. The corresponding electric-field distribution for a three-layer graded system is shown in Fig. 4.9, demonstrating how the stepped permittivity profile produces a more balanced and controlled stress distribution across the insulation thickness.

Another approach to stress control is intersheath grading, which involves placing metallic sheaths at predetermined intermediate potentials within the insulation.

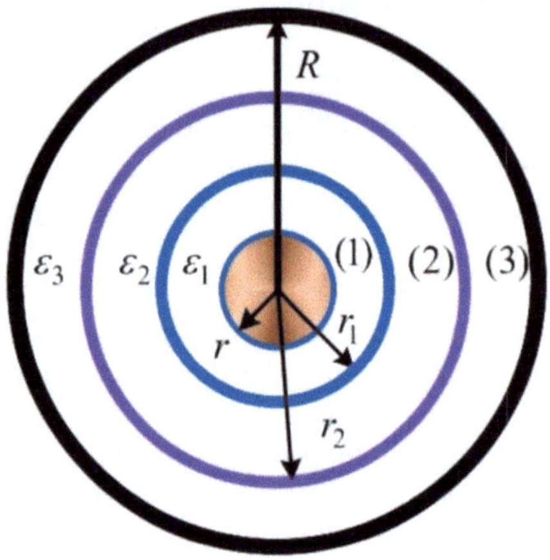

Fig. 4.8 Capacitive grading in a multi-layer dielectric cable

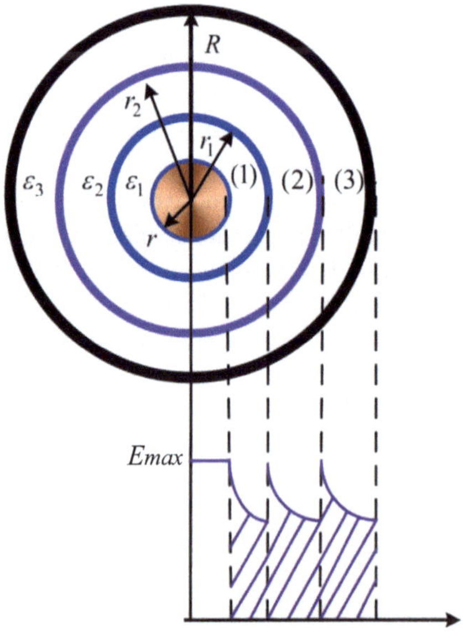

Fig. 4.9 Resulting field distribution in a three-layer graded insulation system

This method allows even finer control of electric-field gradients, as the potential across each section of insulation can be independently regulated. Although inter-sheath grading can achieve near-ideal stress uniformity, it increases construction complexity and requires precise voltage control, making it less common in modern cable systems compared to capacitive grading.

Overall, dielectric grading plays a critical role in high-voltage cable engineering. By flattening the electric-field profile, grading reduces peak stress, prevents premature insulation aging, minimises material usage, and enhances the reliability and operational lifespan of power cables.

4.4.6 Losses in Cables

In power cables, heat generation arises primarily from three distinct types of losses: copper loss, dielectric loss, and intersheath loss. These losses contribute to temperature rise within the cable, influencing its current-carrying capacity, insulation life, and overall operational reliability. Understanding each loss mechanism is essential for accurate thermal modelling and for designing cables that operate safely within permissible temperature limits.

Copper loss is the most fundamental source of heating and is associated with the DC and AC resistance of the conductor. As current flows through the conductor, energy is dissipated in the form of heat due to the inherent resistance of the metallic path. This loss is proportional to the square of the current and is given by:

$$P_{\mathrm{cu}} = I^2 R_c \tag{4.33}$$

where, I is the current flowing through the cable, R_c is the conductor resistance per unit length.

During high-load operation or under fault conditions, copper loss becomes the dominant heating factor, imposing constraints on the permissible operating temperature of the conductor.

A power cable behaves electrically like a cylindrical capacitor, where the conductor serves as one plate and the metallic sheath acts as the other. These plates are separated by the insulation material, which ideally should be a perfect dielectric. However, practical dielectric materials allow small leakage currents and exhibit polarisation effects under alternating electric fields, leading to dielectric loss.

The leakage-current component flowing through the insulation resistance contributes to power dissipation in the dielectric. This component can be expressed as:

$$P_{\mathrm{lc}} = \frac{V^2}{R} \tag{4.34}$$

where V is the applied voltage to a parallel RC circuit and R is the insulation resistance.

More comprehensively, dielectric loss in AC cables is characterised using the loss tangent, $\tan\delta$, which expresses the phase relationship between the resistive leakage current I_r and capacitive current I_c:

$$\tan\delta = \frac{I_r}{I_c} \tag{4.35}$$

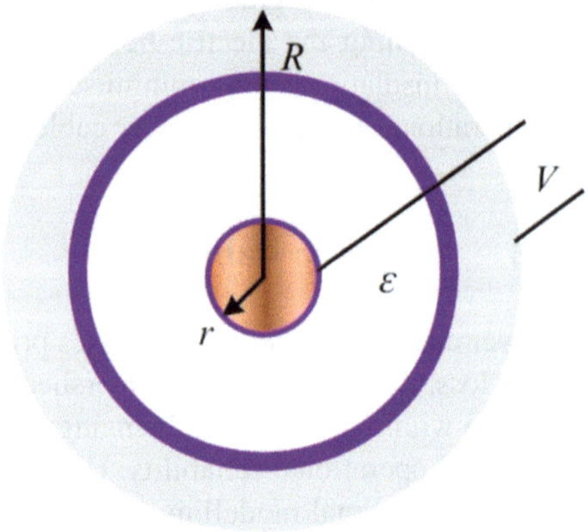

Fig. 4.10 Cross-section of a single-core cable

Using this parameter, the dielectric loss per phase becomes:

$$P_{lc} = V^2 \omega C \tan \delta \tag{4.36}$$

where, ω is the angular frequency, and C is the capacitance between the conductor and the sheath.

Because the capacitance depends on the geometry of the cable, specifically the conductor radius r and sheath radius R cables operating at higher voltages or with thicker insulation tend to exhibit more pronounced dielectric losses. A cross-section of a typical single-core cable showing these geometric parameters is illustrated in the corresponding Fig. 4.10.

In cables equipped with multiple metallic sheaths or screens, circulating currents can develop between the sheaths due to induced voltages or potential differences. These circulating currents produce intersheath losses, which manifest as additional heating in the metallic layers surrounding the insulation. Although intersheath configurations can be used for stress grading in high-voltage systems, improper design may increase losses and reduce overall efficiency.

4.5 PowerFactory Modelling

4.5.1 Modelling of Transmission Systems and Tower

In PowerFactory, overhead transmission lines and cables can be modelled accurately for both steady-state and transient studies. As a starting point, a single machine power system can be modelled as shown in Fig. 4.11. The rated voltage is 11 kV.

After connecting all components by double-clicking on the synchronous generator and selecting the type of 28 MVA GT from the DIgSILENT library, the modelling of the synchronous generator will be completed, as shown in Fig. 4.12.

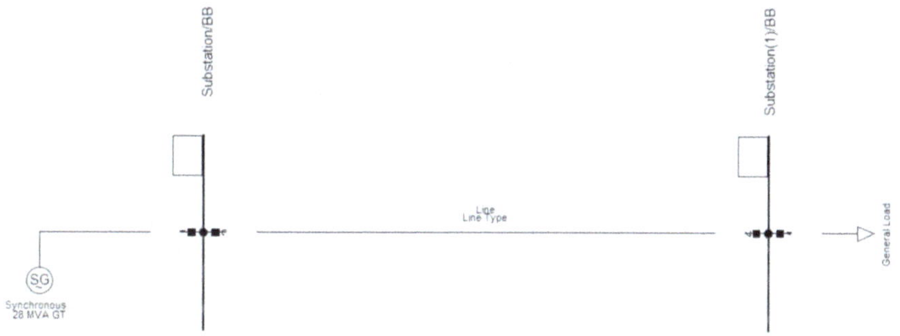

Fig. 4.11 Single-machine power system

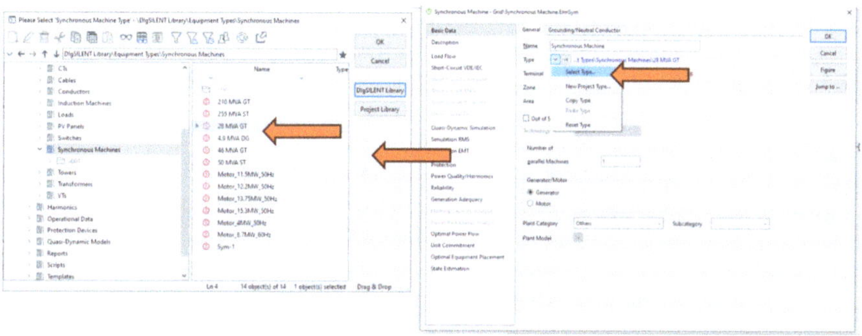

Fig. 4.12 Synchronous generator setting

The load details can be inserted into the general load model, as shown in Fig. 4.13. The active power is 2 MW.

To start the modelling of overhead lines in PowerFactory, double-click on the line, and click on the New Project Type and Tower Type, as shown in Fig. 4.14. Just make sure the length of the line is set as 50 km.

In the new window, the tower can be defined. In the general tap, the number of wires and circuits can be defined. In the geometry tab, the geometry of conductors and towers can be defined. In conductor types of line circuit, by clicking on circuit 1, as shown in Fig. 4.15, a new window for determining the conductors will be opened, as shown in Fig. 4.16, where, by clicking on the new object, the conductor characteristics can be entered, as shown in Fig. 4.17. Since the circuit is at 11 kV, the rated voltage should be changed accordingly. The skin effect can also be checked or unchecked in this window. Checking the skin effect will increase the resistance of the conductor.

By clicking OK from this window and returning to the main window of tower type, as shown in Fig. 4.18, the geometry tab should be selected. In this tab,

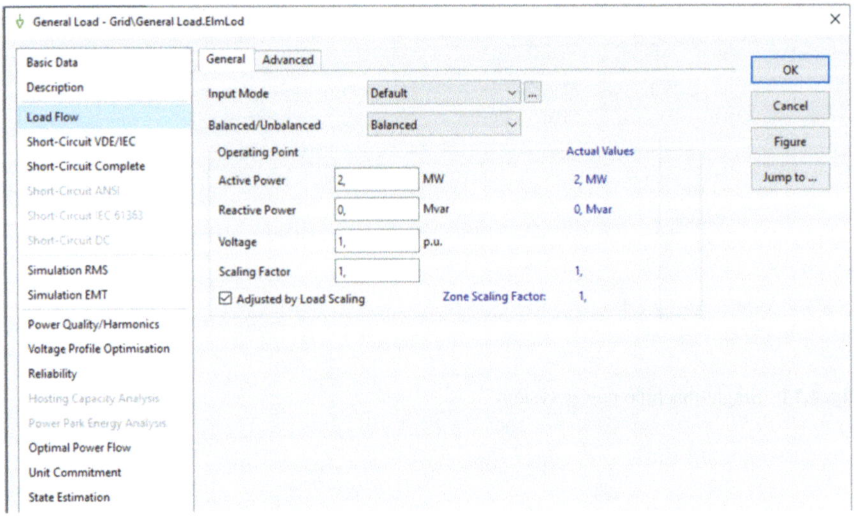

Fig. 4.13 Load setting

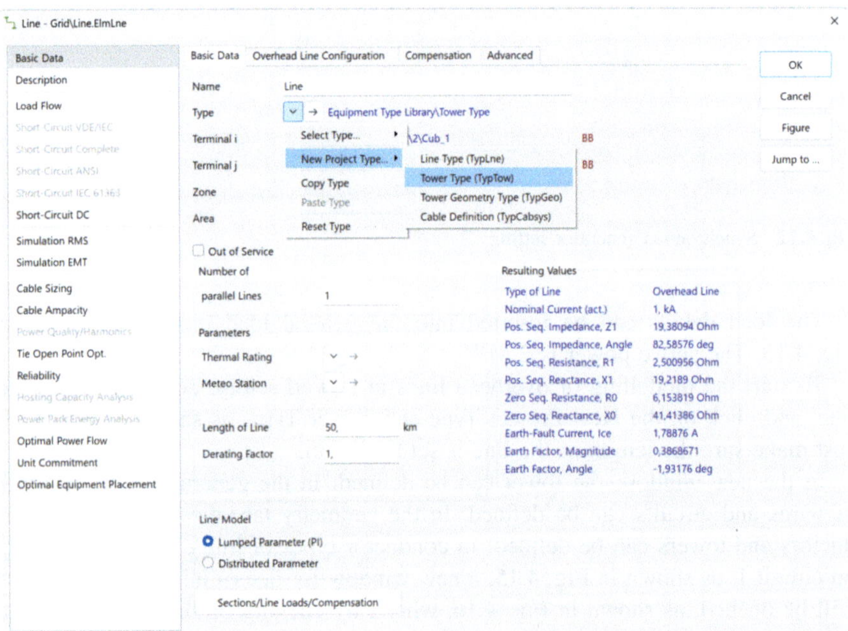

Fig. 4.14 Main window of lines

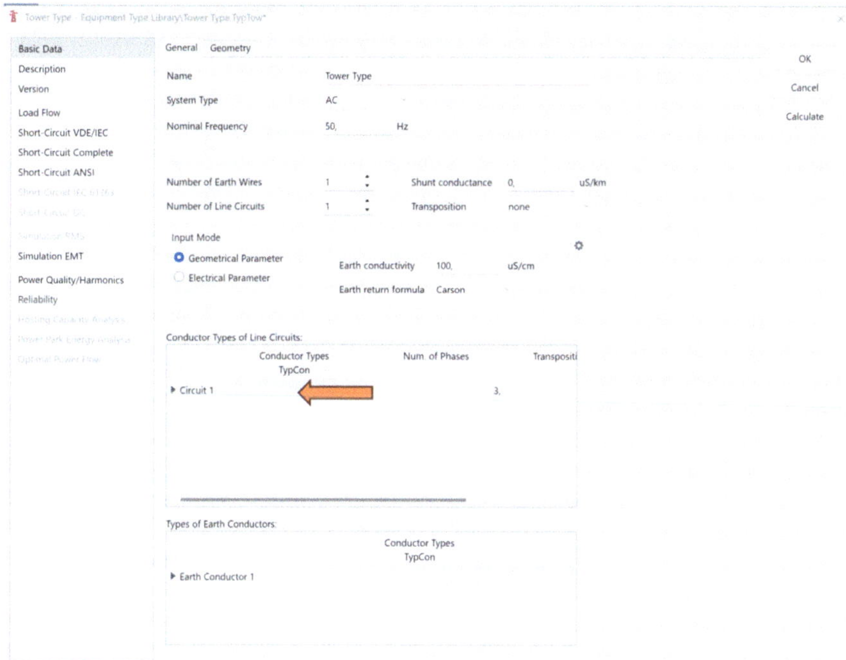

Fig. 4.15 Tower type

Fig. 4.16 Add conductor to tower type

the location of each conductor can be defined by x and y axes coordinates. For example, as shown in Fig. 4.19, the tower is at $x = 0$. By inserting the coordinate as shown in this figure, the coordinates of phases 1, 2, and 3 will be (−2, 17), (0, 19), and (4, 14), respectively. The earth conductor is also located at (0,15). As can be seen from the geometry diagram, the conductors are not symmetrical, which will cause unbalancing in the system.

To observe the voltage and current vector plots, right-click on the line, and selecting plots, insert vector plot, and then current/voltage, as shown in Fig. 4.20,

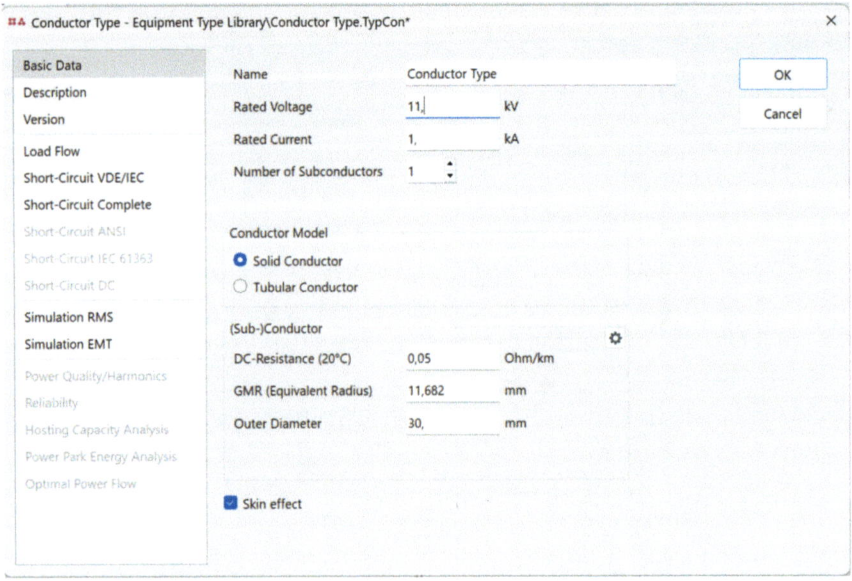

Fig. 4.17 Conductor type

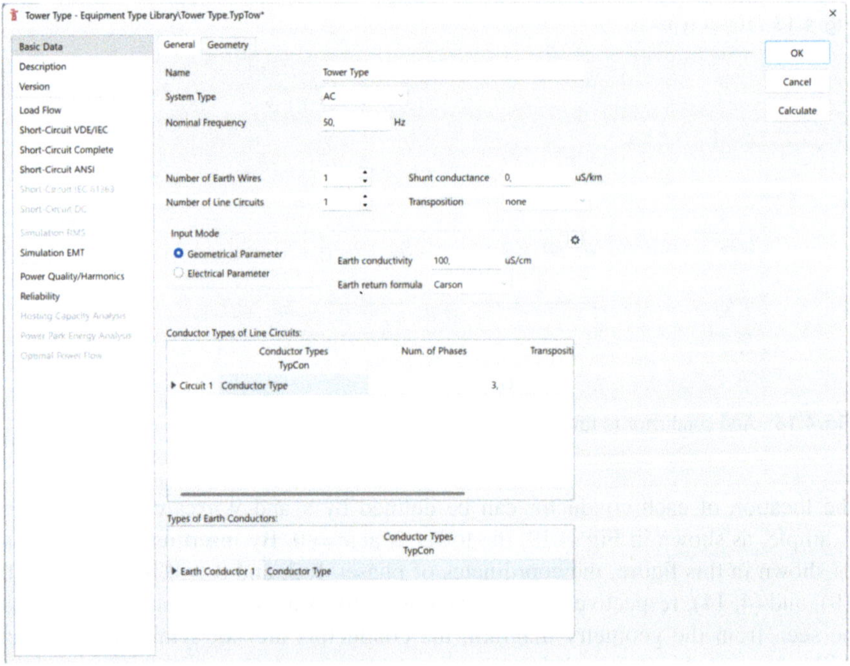

Fig. 4.18 Tower type including conductor

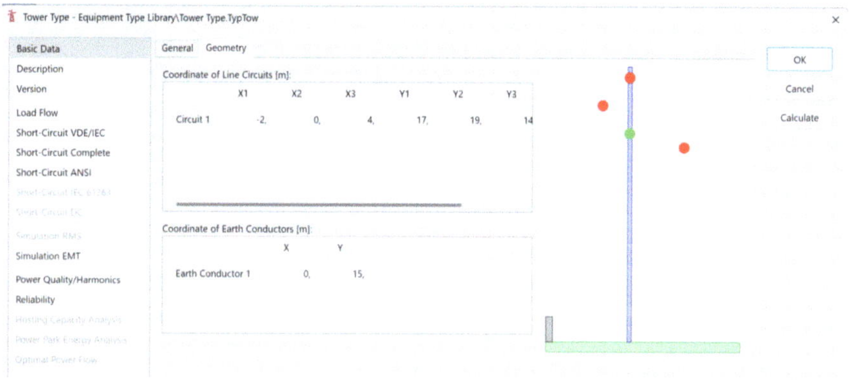

Fig. 4.19 Tower geometry design

and a new plot will be opened as shown in Fig. 4.21. In this figure, it can be seen that the voltage and current are not balanced due to the geometry of the conductors. In the top plot in Fig. 4.21, the sending end voltage is shown, which is balanced, as it is from the generator voltage. However, in the down plot, the receiving end voltage can be seen. In addition to the angle unbalancing, more or less than 12000 differences, the magnitude of voltages is also different.

To solve the unbalancing in the voltages due to the geometry of the tower, the main practical solution is to use transposition in the lines, which can be selected from the main window of tower type, as shown in Fig. 4.22. During the transposition, the order of lines will be changed every couple of kilometers to compensate for the unbalancing in the lines. After applying the transposition, the voltage and current will be balanced, as shown in Fig. 4.23.

4.5.2 Modelling of Cables

In distribution grids, typically, underground cables have been used as transmission lines. To model these cables, in PowerFactory, double-click on the line, and make a new type of Cable definition, as shown in Fig. 4.24. Then, the new window of cable definition will be opened, as shown in Fig. 4.25. In this new window, the circuit position and conductor circuits can be inserted. By double-clicking in the circuit 1 cable type, and making a new object, the type of core can be selected as shown in Fig. 4.26. In this example, we use multicore cables.

In the new window, in the tab of pipe definition, the radius of cable, insulators, etc., could be defined, as shown in Fig. 4.27. Then, in the conductor coordination, the positions of the conductors in the pipe could be defined as shown in Fig. 4.28. The coordinates for conductors can be defined based on the angle from the center and their distance from it.

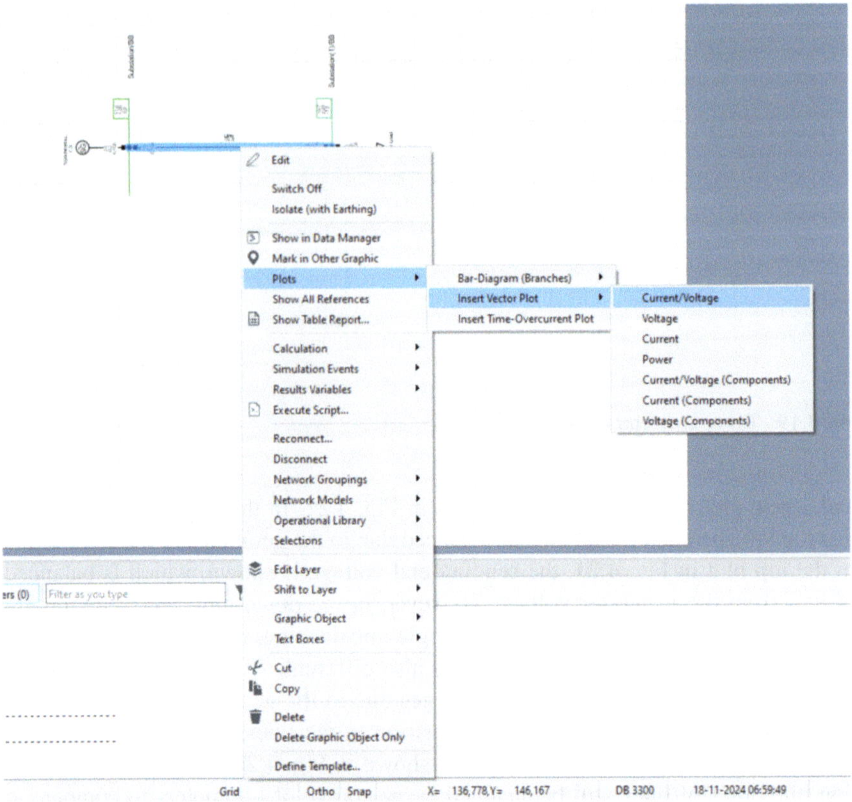

Fig. 4.20 Plotting current/voltage vector

By clicking ok and returning to the main window of cable type, the material, thickness, and details for each conductor can be defined, as shown in Fig. 4.29.

After defining the cable, the details of the cable can be defined through the cable ampacity calculations through the PowerFactory, by right-clicking on the line, and selecting calculations, and cable ampacity calculations, as shown in Fig. 4.30.

Therefore, in the new window, by selecting the calculation method based on IEC 60287 and executing it, the calculation will be performed, as shown in Fig. 4.31. To show the report, by selecting the cable analysis toolbox, as shown in Fig. 4.32, and executing the cable ampacity report, the detailed report of the cable, like losses, maximum current, and thermal limitation will be shown, as presented in Table 4.2.

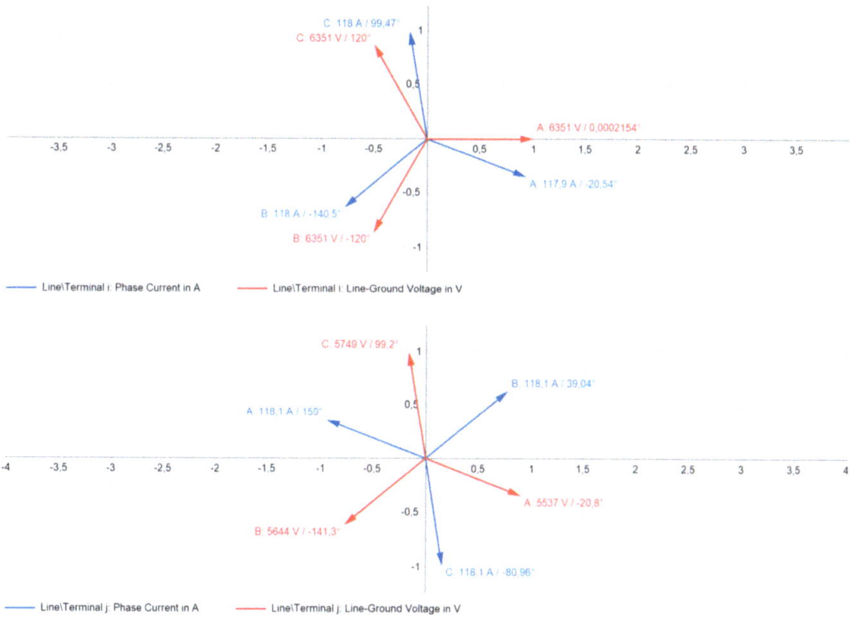

Fig. 4.21 Unbalanced current/voltage vector plot

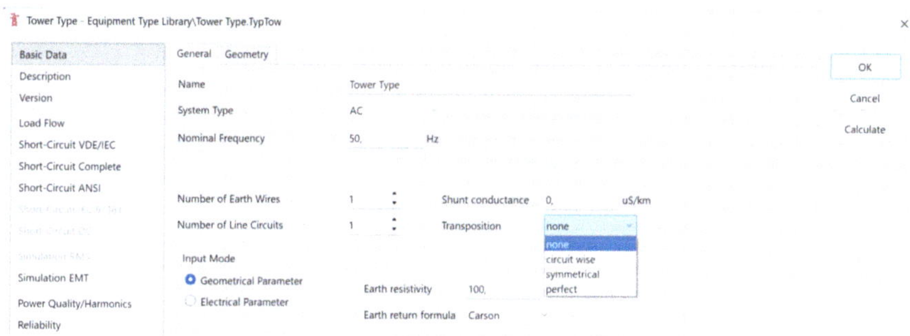

Fig. 4.22 Transposition in towers

4.6 Summary

This chapter has provided a comprehensive treatment of overhead transmission lines and underground cables, the essential links that transport electrical energy across power systems. It began by detailing the physical parameters of lines—series resistance, inductance, and shunt capacitance—including their calculation for single-circuit, bundled-conductor, and transposed configurations. Transmission

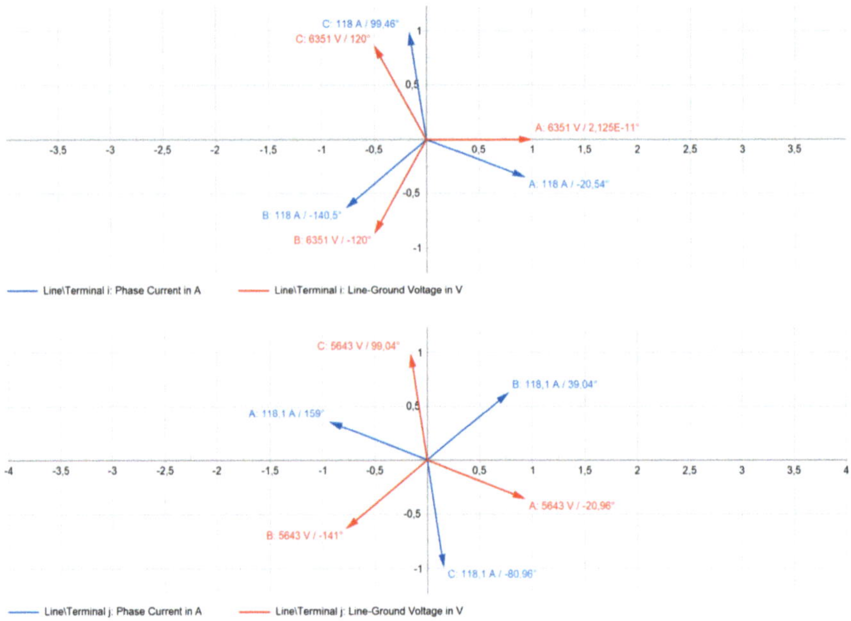

Fig. 4.23 Balance current/voltage vectors with perfect transposition

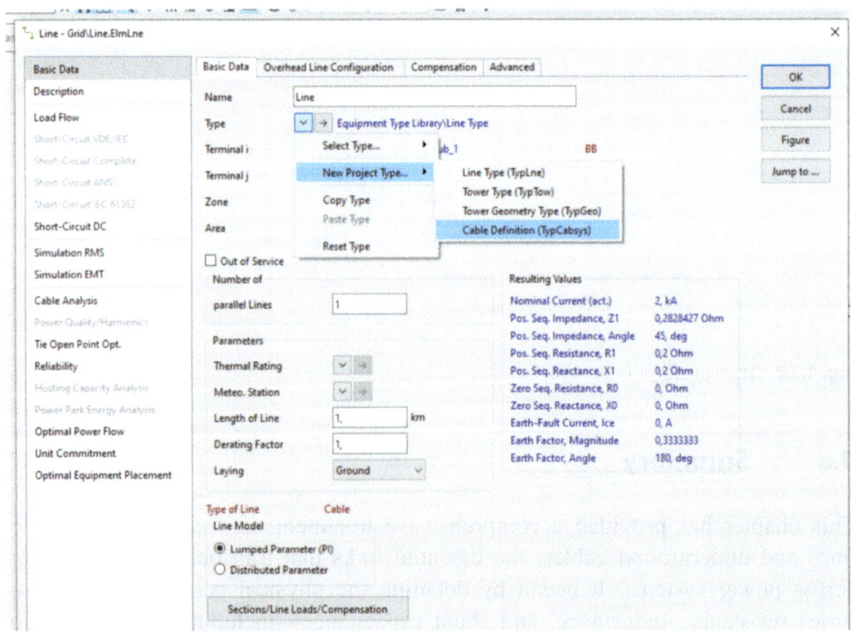

Fig. 4.24 Line window for inserting cable

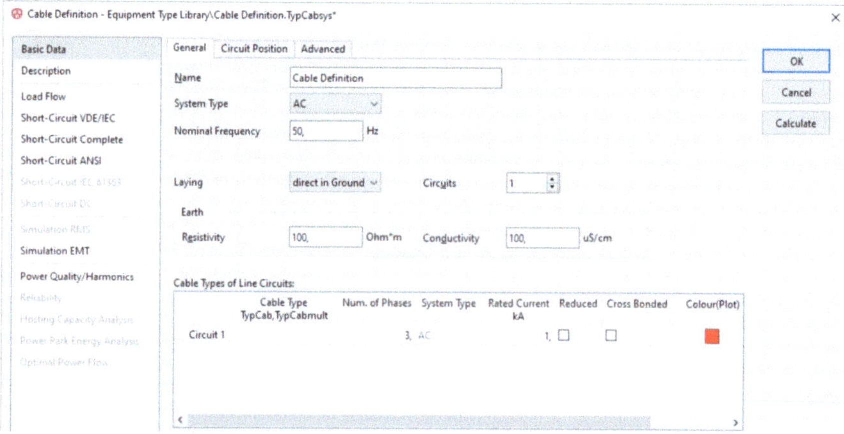

Fig. 4.25 Cable definition

Fig. 4.26 Adding multicore/pipe cable

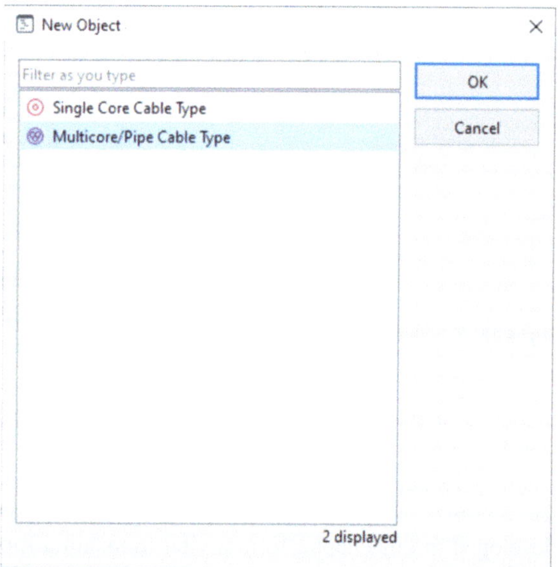

lines were classified as short, medium, and long, with corresponding lumped and distributed equivalent-circuit models (nominal-π, nominal-T, and hyperbolic ABCD-parameter representations) derived for accurate performance evaluation. Key operational phenomena such as voltage drop, power losses, surge-impedance loading, Ferranti effect, voltage regulation, corona discharge, and skin effect were examined together with their impacts on efficiency, stability, and power quality. The distinctive features of cables (higher capacitance, lower inductance, and special insulation requirements) were contrasted with overhead lines, highlighting

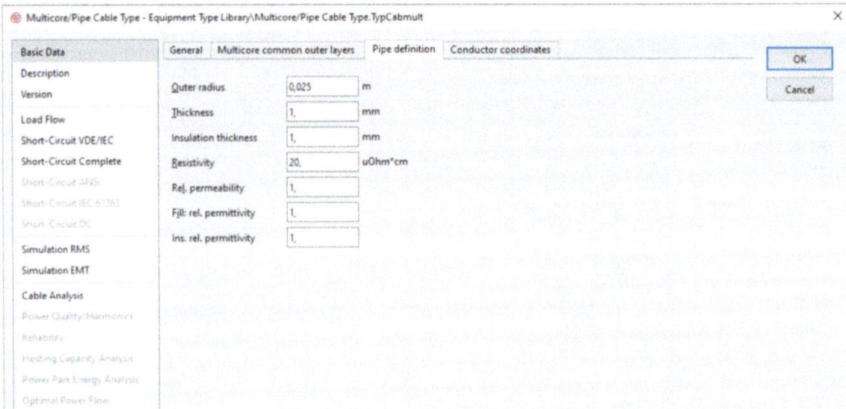

Fig. 4.27 Pipe definition

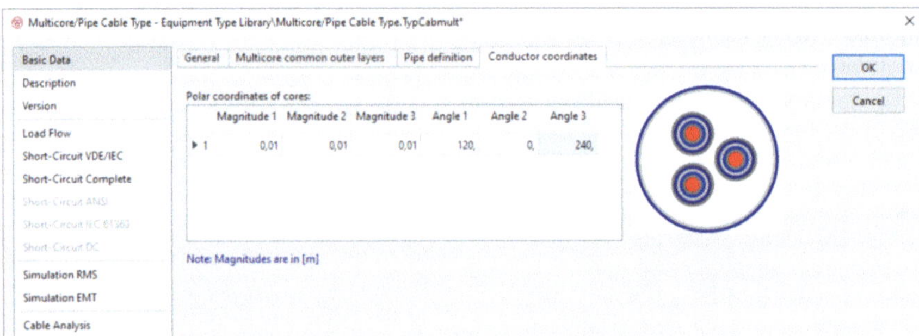

Fig. 4.28 Conductor coordinates for cables

modeling challenges for underground and submarine applications. Finally, detailed step-by-step modeling tutorials in PowerFactory—DIgSILENT were presented, enabling readers to build, parameterize, and analyze realistic line and cable models for load-flow, short-circuit, and dynamic studies. The concepts and practical skills developed in this chapter form a vital foundation for the power-flow, short-circuit, protection, and stability analyses covered in the following chapters.

4.7 Problems and Questions

1. A 220 kV, 50 Hz, 150 km long three-phase overhead transmission line has the following per-kilometer parameters: $R = 0.12\,\Omega/\text{km}$, $L = 1.25\,\text{mH/km}$, $C = 0.009\,\mu\text{F/km}$. Classify the line as short, medium, or long and calculate (a) the total series impedance Z and shunt admittance Y referred to the nominal-π

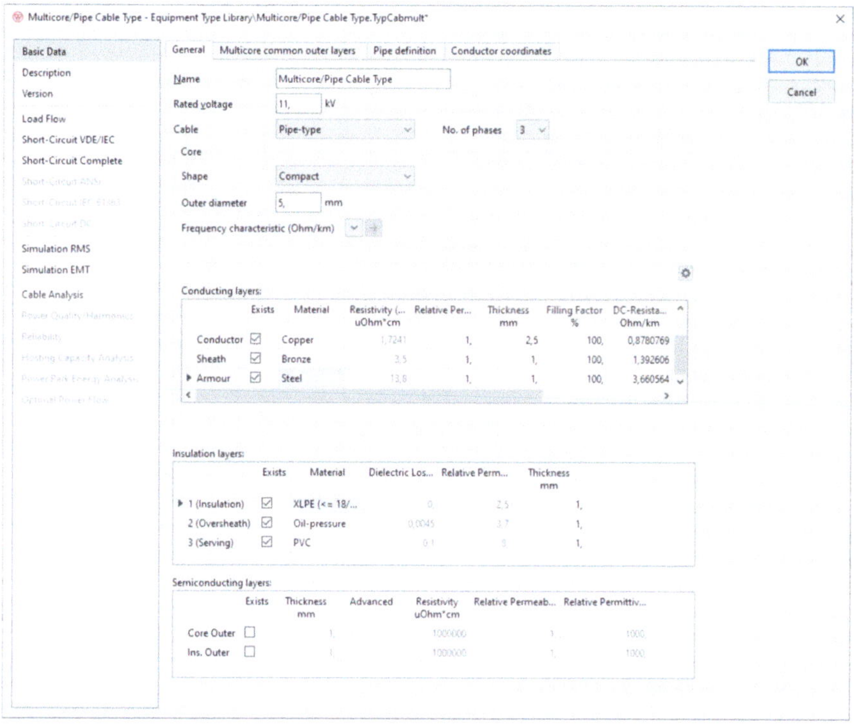

Fig. 4.29 General window of pipe cable type definition

model, (b) the voltage regulation when the line delivers 200 MW at 0.9 lagging power factor with the receiving-end voltage held at 220 kV, and (c) the surge-impedance loading (SIL) of the line.

2. An underground cable circuit operates at 132 kV and has a total length of 25 km with series resistance $R = 0.08\,\Omega/\text{km}$, inductance $L = 0.35\,\text{mH/km}$, and capacitance $C = 0.25\,\mu\text{F/km}$. Using the nominal-π model, determine (a) the sending-end voltage and (b) the total three-phase active and reactive power losses when the cable supplies a 120 MVA load at 0.85 lagging power factor with the receiving-end voltage maintained at rated value. Compare the capacitive charging current with that of an equivalent overhead line and explain why cable lines are more prone to the Ferranti effect.

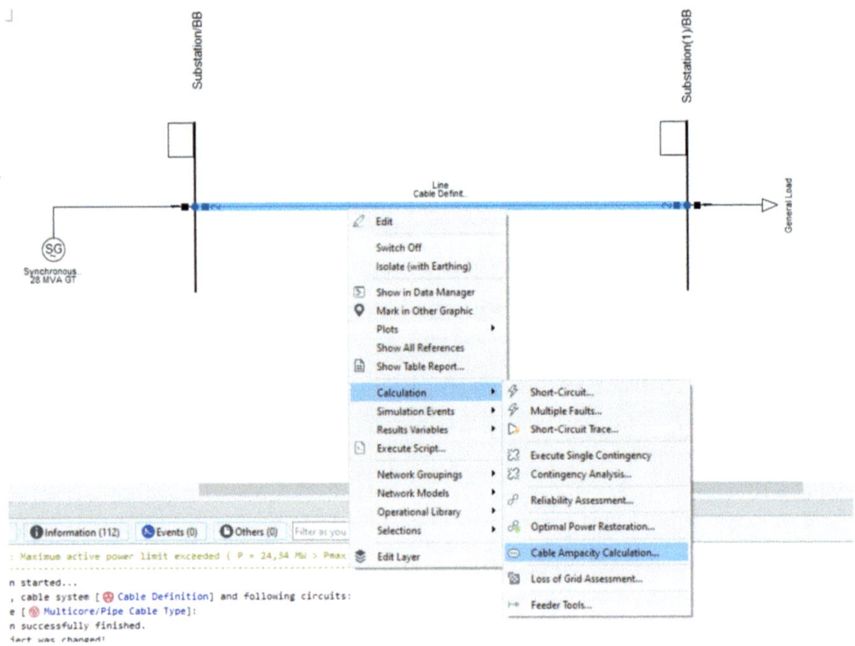

Fig. 4.30 Cable ampacity calculation

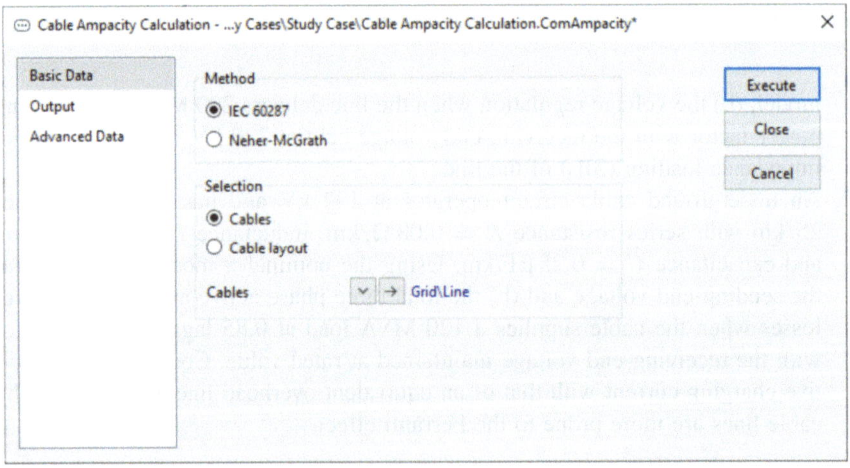

Fig. 4.31 Cable ampacity calculation window

Fig. 4.32 Cable analysis report

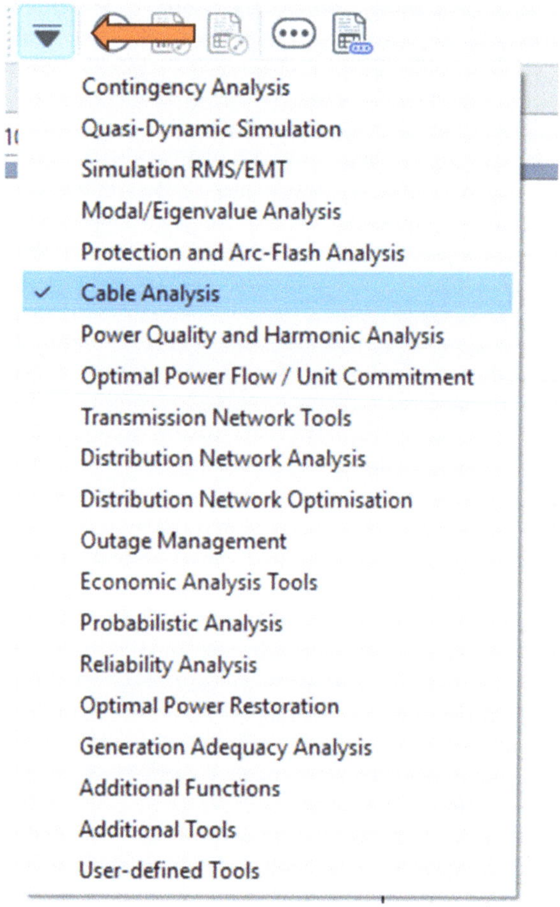

Table 4.2 Cable ampacity report: cable ampacity

Study case		Study case					
Result file		Cable ampacity results					
Used method		IEC 60287					
Line/ line section	Cable	Max. current kA	Thermal resistance (conductor and sheath) K m/W	Loss factor for armour or pipe	Dielectric losses W/ m	Conductor losses W/ m	Temperature of medium °C
Line	Multicore/ pipe cable type	0,100615	0,053551	0,001291	0,125529	10,986755	42,499070

Power Flow and Short Circuit Calculations

5

Abstract

This chapter has provided a complete theoretical and practical treatment of power-flow and short-circuit calculations, the two core computational tools used for steady-state and fault analysis in modern power systems. It began by formulating the nonlinear power-balance equations, classifying buses as slack, PV, and PQ, and deriving the solution algorithms (Gauss-Seidel, Newton-Raphson, and fast decoupled methods) together with convergence criteria and practical considerations for large-scale networks. The chapter then developed symmetrical-component theory for short-circuit studies, deriving fault-current expressions for the four major fault types (three-phase, single-line-to-ground, line-to-line, and double-line-to-ground) and explaining their application in equipment rating, protection settings, and system design. Finally, detailed step-by-step tutorials in PowerFactory—DIgSILENT demonstrated how to execute balanced and unbalanced AC load-flow studies, steady-state short-circuit calculations with user-defined fault locations and resistances, and transient RMS/EMT simulations using predefined events, result variables, and waveform plotting. The analytical methods and software skills presented in this chapter form the essential bridge to the protection and stability studies covered in Chaps. 6 and 7.

Keywords

Power flow • Short-circuit calculations • Symmetrical components

5.1 Introduction

Load flow, or power flow, study is a fundamental analytical tool used in the planning, operation, maintenance, and control of electrical power systems. It provides essential information about the steady-state behavior of a power network, enabling

N. Bayati and M. A. Khan, *Power System Fundamentals with PowerFactory*,
https://doi.org/10.1007/978-3-032-32450-4_5

engineers to understand how electric power moves from generation sources to consumers through transmission and distribution systems.

During the planning stage, load flow studies help determine whether particular elements of the power system are underloaded or overloaded. Such analysis assists in identifying the need for capacity expansion, network reinforcement, or equipment upgrades. By predicting how anticipated load growth and system changes will affect performance, planners can ensure that the power system remains robust and reliable over time.

In operational studies, load flow analysis plays a key role in verifying that each generator operates at or near its optimal point. This ensures efficient generation dispatch while maintaining system stability and desired voltage levels. Operators rely on load flow results to balance real and reactive power, avoid excessive line loading, and prevent voltage violations.

In a power system, electricity flows from the sending end to the receiving end through transmission lines. The mathematical relationships describing these real and reactive power flows are known as power flow equations. These equations are inherently nonlinear due to the sinusoidal nature of alternating-current systems. As a result, they must be solved using numerical, iterative methods such as the Newton–Raphson, Gauss–Seidel, or Fast Decoupled load flow algorithms.

A typical load flow study calculates a variety of important system parameters. These include the voltage magnitude and phase angle at each bus, real and reactive power flowing through all branches, voltage drops across feeders, and total system power losses. Such comprehensive information is vital for ensuring reliable system operation, maintaining power quality, and planning future network development.

5.2 Classification of Buses

In power system load flow analysis, buses serve as fundamental reference points that describe the operating state of each node within the network. A "bus" represents a junction where generators, loads, and transmission lines are interconnected. To solve the load flow equations efficiently, buses are categorized into three principal types: Load (PQ) buses, Generator (PV) buses, and Slack (reference) buses. Each bus type specifies certain electrical quantities while requiring others to be determined through iterative computation. Understanding the characteristics and roles of these buses is essential for accurate modeling and analysis of large-scale power systems.

A load bus, commonly referred to as a PQ bus, is a bus in which the real power (P) and reactive power (Q) are specified. These quantities represent the power consumed by the load at that bus. The voltage magnitude (V) and the phase angle δ are not known beforehand and must be determined through the load flow study. In this convention, real and reactive power supplied to the system are treated as positive values, while power consumed from the system is considered negative.

A generator bus, also known as a voltage-controlled or regulated bus, is characterized by a specified real power output P and a controlled V. For this reason,

Table 5.1 Different types of bus bars in power system studies

Type of buses	Known parameter	Unknown parameter
Slack bus	V, δ	P, Q
Load bus	P, Q	|V|, δ
Generator bus	P, |V|	Q, δ

it is commonly called a PV bus. The generator adjusts its reactive power Q to maintain the specified voltage, and both the Q and the δ are unknown and must be computed during the load flow analysis.

The slack bus, often called the swing bus or reference bus, is the bus where both the voltage magnitude and phase angle are specified. Typically, this bus is assigned a voltage magnitude of 1.0 per unit and a phase angle of zero degrees, serving as the reference point for all other angles in the system. The P and Q at the slack bus are not specified and must be calculated by the load flow program to balance the total system power and account for losses. A summary of each bus bar is described in Table 5.1.

5.3 Power Flow in Two-Bus Systems

The two-bus system represents the simplest possible structure used to study power flow in electrical power networks. Despite its simplicity, it provides valuable insight into the fundamental relationships between voltage magnitudes, voltage phase angles, line impedance, and the resulting real and reactive power flows. Because larger power networks are essentially composed of interconnected buses linked through transmission lines, understanding the two-bus model serves as an essential foundation for analyzing more complex multi-bus systems.

In a typical two-bus system, one bus acts as a sending-end source (V_s), often modeled as a generator or an infinite bus, while the other acts as a receiving-end load bus (V_r), as shown in Fig. 5.1. The two buses are connected through a transmission line with known electrical parameters such as resistance, reactance, and, in some cases, shunt admittance. By applying KVL to the circuit V_s can be calculated.

$$V_s = V_r + I(R + jX_L) \tag{5.1}$$

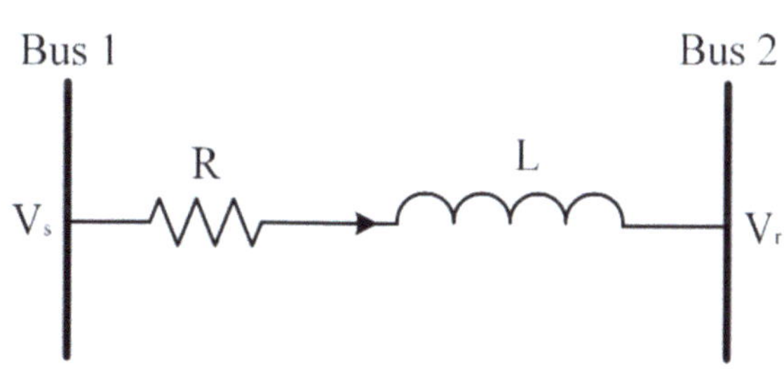

Fig. 5.1 Layout for two bus system

The value of the resistance of the transmission line is very small, as compared to inductive reactance. As a result, we obtain:

$$I = \frac{V_s - V_r}{jX_L} \tag{5.2}$$

$$I_s^* = I^* = I_r^* = \left(\frac{V_s - V_r}{jX_L}\right)^* \tag{5.3}$$

Power flows from the sending bus to the receiving bus depending on the voltage difference and the phase angle difference between the two buses. This relationship illustrates one of the fundamental principles of AC power systems: real power flow is primarily driven by phase angle differences, while reactive power flow is influenced by voltage magnitude differences.

$$P_r = \frac{V_s V_r}{X_L} \sin\delta \tag{5.4}$$

$$Q_r = \frac{V_s V_r}{X_L} \cos\delta - \frac{V_r^2}{X_L} \tag{5.5}$$

Because the sine function reaches its maximum value when $\delta = 90°$, the real power delivered at both ends of the line is maximized at this operating condition. This theoretical limit is known as the maximum power transfer capability of the line. Substituting $\sin 90° = 1$ into the power equation gives the maximum power expression

$$P_{\max} = \frac{V_1 V_2}{X} \tag{5.6}$$

This result provides important insight into transmission line design and operational stability. Although the mathematical maximum occurs at 90°, practical systems never operate near this limit because system stability margins must be maintained. Operating at or close to $\delta = 90°$ would leave the power system extremely vulnerable to disturbances. Therefore, the maximum real power expression is primarily used as a theoretical benchmark to evaluate system strength, stability, and transmission capability.

The average reactive power flow through the line is expressed as:

$$Q_{\text{avg}} = \frac{Q_s + Q_r}{2} \tag{5.7}$$

$$Q_{\text{avg}} = \frac{1}{2X_L}\left(\mathrm{V_s^2} - \mathrm{V_r^2}\right) \tag{5.8}$$

In summary, the two-bus power transfer model helps build the basic understanding of how voltage magnitudes, line reactance, and the phase angle determine the flow of real and reactive power between two points in a power system. It shows clearly how power increases with the torque angle and reaches a theoretical limit at ninety degrees. While this simple arrangement is helpful for grasping core ideas, real networks involve more buses and more complex interactions. Moving from a two-bus system to a three-bus system introduces additional paths for power flow, new angle relationships, and more realistic network behaviour. This transition marks an important step toward understanding how large, interconnected power systems operate and how power is shared among multiple buses.

5.4 Power Flow in Three-Bus Systems

A three-bus system provides the simplest network that captures the essential characteristics of real power flow behaviour while remaining small enough for analytical and computational demonstration. Unlike the two-bus model, a three-bus configuration introduces multiple transmission paths, power sharing between generators, and interactions between generation and load through interconnected lines. For this reason, it is the most widely used teaching model in power systems courses and textbooks.

Figure 5.2 shows the basic topology of the three-bus network. Generator G1 injects current I_1 into Bus 1, generator G2 injects current I_2 into Bus 2, and a load draws current I_3 from Bus 3. The buses are interconnected by three transmission lines of impedances Z_{12}, Z_{23}, and Z_{31}. Power injected at any bus can therefore flow through multiple paths to reach the load, depending on the relative values of line impedances. This multi-path characteristic is fundamental to understanding loop flows, voltage coupling, and loss distribution in larger power networks.

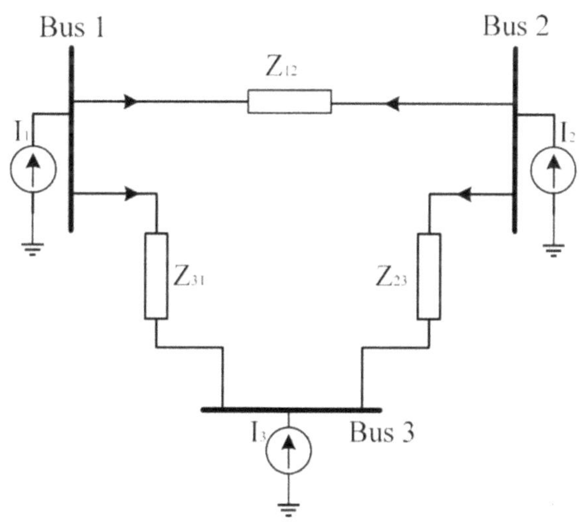

Fig. 5.2 Three-bus system with line impedances Z_{12}, Z_{23}, Z_{31}, and injected currents I_1, I_2, I_3

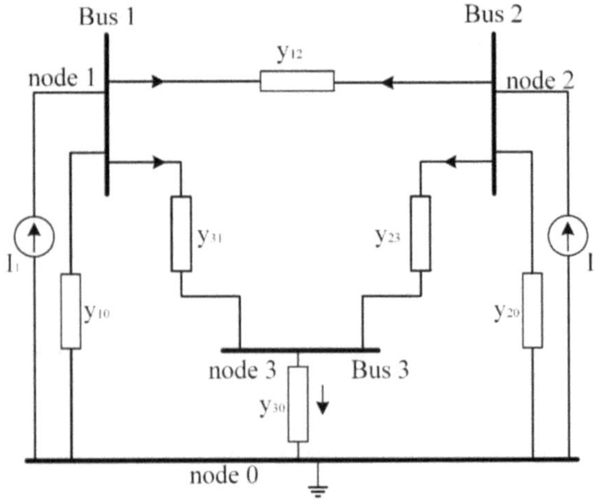

Fig. 5.3 Admittance representation showing self-admittances y_{10}, y_{20}, y_{30} and mutual admittances y_{12}, y_{23}, y_{31}

To analyse this network mathematically, the impedance diagram is converted into its corresponding admittance representation. Figure 5.3 illustrates the equivalent admittance model where each line is represented by its admittance $y_{mn} = 1/Z_{mn}$. . Shunt admittances to the reference node (node 0) may also be included depending online charging and grounding effects. Each node is connected to the reference node via the self-admittances y_{10}, y_{20}, and y_{30}, while the mutual admittances y_{12}, y_{23}, and y_{31} represent the coupling between buses.

Using this representation, the network can be modelled compactly through the bus admittance matrix (Y-bus), relating bus voltages to bus injection currents:

$$\begin{bmatrix} Y_{11} & Y_{12} & Y_{13} \\ Y_{21} & Y_{22} & Y_{23} \\ Y_{31} & Y_{32} & Y_{33} \end{bmatrix} \begin{bmatrix} V_1 \\ V_2 \\ V_3 \end{bmatrix} = \begin{bmatrix} I_1 \\ I_2 \\ I_3 \end{bmatrix} \tag{5.9}$$

The diagonal elements Y_{ii} represent the total admittance connected to bus i, and the off-diagonal terms $Y_{ij} = -y_{ij}$ represent the negative of the admittance linking buses i and j. This Y-bus formulation is the foundation for all load flow methods, Gauss–Seidel, Newton–Raphson, and Fast Decoupled Load Flow.

Since the injected complex power at a bus is defined as $S_i = P_i + jQ_i = V_i I_i^*$, substituting into Eq. (5.9) gives an iterative expression for solving unknown voltages:

$$V_i = \frac{1}{Y_{ii}} \left[\frac{P_i - jQ_i}{V_i^*} - \sum_{\substack{j=1 \\ j \neq i}}^{N} Y_{ij} V_j \right] \tag{5.10}$$

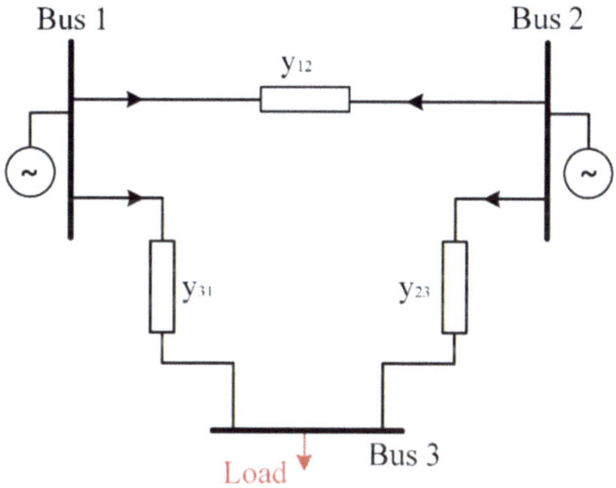

Fig. 5.4 A practical three-bus single-line diagram with generators G1 and G2 supplying a load at Bus 3

This equation is applied repeatedly during iterative load flow computation. The three-bus system is exceptionally useful here because each bus typically represents a different operating category:

Bus 1 (Slack Bus): Voltage magnitude and angle are specified; this bus absorbs the numerical mismatch between calculated and specified power. It represents generator G1.

Bus 2 (PV Bus): Active power P_2 and voltage magnitude $|V_2|$ are specified. Generator G2 controls the voltage at this bus through reactive power support.

Bus 3 (PQ Bus): Represents the load with specified real and reactive power demand. Its voltage magnitude and angle are unknown and must be computed.

Figure 5.4 shows the simplified single-line diagram corresponding to this bus configuration. Generators G1 and G2 supply the network through lines with admittances y_{12}, y_{23}, and y_{31}, while Bus 3 connects to the load.

Because power can flow from Bus 1 to Bus 3 either directly through line y_{31} or via Bus 2 through $y_{12}$and y_{23}, the network naturally exhibits loop flows and voltage dependencies. An increase in load at Bus 3 affects voltages at all buses, and generator reactive power limits at Bus 2 may cause the PV bus to temporarily switch to PQ mode during calculations, mirroring behaviour seen in real transmission systems.

For these reasons, the three-bus system remains an indispensable benchmark model, providing insight into voltage control, reactive power distribution, losses, line loading, and iterative convergence behaviour in modern power flow studies.

5.5 Power Flow Methods

Power flow analysis in electrical power systems requires solving nonlinear algebraic equations that relate real and reactive power injections to bus voltages and network admittances. Because these equations incorporate trigonometric functions and products of voltage magnitudes and angles, analytical closed-form solutions are not feasible for systems of practical size. Instead, numerical iterative methods are used to determine a steady-state operating condition that satisfies network constraints. Among these numerical approaches, the Gauss–Seidel method, the Newton–Raphson method, and the Fast Decoupled Load Flow method have become the principal tools used in both academic study and industrial system simulations. Each of these methods employs a different strategy to deal with the nonlinearity of power flow equations, and each provides a different compromise between computational effort, convergence speed, and robustness.

5.5.1 Gauss–Seidel Method

The Gauss–Seidel method is one of the earliest and simplest iterative approaches for solving load flow problems. The method updates one bus voltage at a time by solving the complex power equation for that bus and immediately substitutes the newly computed value into the next calculation. This sequential updating creates a cascading effect where information propagates through the system during each iteration.

The complex power injected at bus i is expressed as

$$S_i = P_i + jQ_i = V_i I_i^* \tag{5.11}$$

and the injected current can be written through the admittance matrix as

$$I_i = \sum_{j=1}^{N} Y_{ij} V_j. \tag{5.12}$$

Substituting this into the power expression yields the Gauss–Seidel voltage update formula for the $(k + 1)$th iteration:

$$V_i^{(k+1)} = \frac{1}{Y_{ii}} \left[\frac{P_i - jQ_i}{\left(V_i^{(k)}\right)^*} - \sum_{\substack{j=1 \\ j \neq i}}^{N} Y_{ij} V_j^{(k+1 \text{ or } k)} \right] \tag{5.13}$$

Convergence is evaluated by comparing successive voltage updates using

$$\Delta V_i = |V_i^{(k+1)} - V_i^{(k)}| \tag{5.14}$$

The method converges more reliably when the diagonal elements of the admittance matrix dominate, which is typical in radial distribution systems. However, in large meshed transmission systems, convergence becomes relatively slow, and in heavily stressed networks the method may fail altogether. Despite these limitations, Gauss–Seidel remains useful for educational demonstrations and small-scale system modelling because of its simplicity and intuitive iterative structure.

5.5.2 Newton–Raphson Method

The Newton–Raphson method is considered the most powerful and widely applicable numerical technique for solving the load flow problem. Instead of updating voltages directly, this method linearises the nonlinear power flow equations about the current operating point using a first-order Taylor expansion. This results in a set of linear equations relating changes in voltage angles and magnitudes to mismatches in real and reactive power.

The active and reactive power mismatches at the bus i are defined as

$$\Delta P_i = P_i^{\text{spec}} - P_i^{\text{calc}}, \Delta Q_i = Q_i^{\text{spec}} - Q_i^{\text{calc}} \tag{5.15}$$

These mismatches form the left-hand side of the Newton–Raphson linearised equation:

$$\begin{bmatrix} \Delta P \\ \Delta Q \end{bmatrix} = \begin{bmatrix} J_1 & J_2 \\ J_3 & J_4 \end{bmatrix} \begin{bmatrix} \Delta \delta \\ \Delta |V| \end{bmatrix}, \tag{5.16}$$

where the Jacobian submatrices consist of partial derivatives,

$$J_1 = \frac{\partial P}{\partial \delta}, J_2 = \frac{\partial P}{\partial |V|}, J_3 = \frac{\partial Q}{\partial \delta}, J_4 = \frac{\partial Q}{\partial |V|} \tag{5.17}$$

Solving this linear system yields the correction terms for voltage angle and magnitude, which are applied iteratively as

$$\delta^{(k+1)} = \delta^{(k)} + \Delta \delta, |V|^{(k+1)} = |V|^{(k)} + \Delta |V|. \tag{5.18}$$

Newton–Raphson exhibits quadratic convergence, meaning the error reduces dramatically with each iteration once the solution is near the correct operating point. This property makes the method especially effective for large power systems with high levels of loading. Although the method requires more computational effort per iteration due to Jacobian formation and matrix inversion, modern computation easily supports these requirements. Consequently, Newton–Raphson is the standard method in most commercial power system analysis packages and is regarded as the most reliable tool for complex and heavily loaded networks.

5.5.3 Fast Decoupled Load Flow Method

The Fast Decoupled Load Flow method is a streamlined variant of the Newton–Raphson method that leverages the physical characteristics of high-voltage transmission lines. In such networks, line reactances dominate over resistances, and the coupling between real power and voltage magnitude, as well as between reactive power and voltage angle, is relatively weak. By exploiting these characteristics, the full Newton–Raphson Jacobian matrix may be simplified and separated into two decoupled subsystems.

Under these assumptions, the power flow equations reduce to the approximate forms

$$\Delta P \approx B' \, \Delta\delta, \, \Delta Q \approx B'' \, \Delta|V| \tag{5.19}$$

where the matrices B' and B'' consist largely of the negative imaginary components of the bus admittance matrix:

$$B' = -\mathrm{Im}(Y_{\mathrm{bus}}), B'' = -\mathrm{Im}(Y_{\mathrm{bus}}) \tag{5.20}$$

The voltage updates are then obtained from

$$\Delta\delta = (B')^{-1}\Delta P, \, \Delta|V| = (B'')^{-1}\Delta Q \tag{5.21}$$

followed by

$$\delta^{(k+1)} = \delta^{(k)} + \Delta\delta, \, |V|^{(k+1)} = |V|^{(k)} + \Delta|V| \tag{5.22}$$

Since the decoupled matrices are constant throughout the iterations, their factorisation need only be performed once. This greatly reduces computational requirements, making the Fast Decoupled method exceptionally efficient for large-scale transmission networks. Although its convergence is not as fast as Newton–Raphson and its accuracy decreases in low-voltage or resistive systems, the method remains widely used in operational planning, contingency analysis, and real-time stability assessments due to its speed, simplicity, and suitability for repeated calculations. All methods are summarized in Table 5.2.

Table 5.2 Summary of power flow calculation methods

Method	Convergence	Computation speed	Best use case
Gauss-Seidel	Slow	Low memory, low speed	Small networks, teaching
Newton-Raphson	Fast (quadratic)	High per iteration, few iterations	Most reliable and accurate
Fast decoupled	Moderate	Extremely fast	Large HV networks, operations

5.6 Short Circuit Calculations

The main objective of power utility companies is to operate the grid under balanced conditions. This normal operating condition of the power system can be disrupted by short-circuit faults due to adverse weather conditions such as heavy wind, lightning strikes, or due to other factors such as birds shorting out the lines, vehicles colliding with transmission lines or towers by accident, or trees falling on the overhead lines. These events trigger the flashover from the resultant high magnitude of the current that passes through the lines to the ground. This condition of the transmission lines is known as a short-circuit or fault. In a faulty situation, a very low impedance path is created between lines to ground, resulting in high magnitude current imposing a heavy duty on the circuit breaker and other controlling equipment.

In distribution and transmission systems, short-circuit faults occur in different configurations with different probabilities. Analysis of historical fault data reveals that single-phase-to-ground faults are by far the most common type of fault, with approximately 65% of all short circuits. Line-to-line faults account for 20%, while double-line-to-ground faults represent 10%. Symmetrical three-phase faults are relatively rare at around 5%. The reasons for these power system faults and resulting outages provide critical insights into the reliability of grids. Weather conditions and vegetation contact emerge as the primary contributors to 30% of the causes. Equipment failures and human errors follow at 26%, with a substantial 20% of incidents having unknown reasons. Vehicle accidents contribute 12%, planned maintenance outages 6%, and animal interference about 5%, with overdemand accounting for less than 1%. These statistics highlight the value of proactive measures such as vegetation management programs, enhanced asset monitoring, and public awareness campaigns to minimize fault occurrences and improve overall grid resilience [1, 2].

In the Baltic and Nordic power systems, short-circuit faults and associated disturbances exhibit consistent patterns across the region. During 2022, the combined Baltic and Nordic areas experienced 1,972 faults, and the country-level variations are pronounced: Finland recorded the highest fault count (596), while Iceland experienced only 85 faults. These statistics illustrate the scale of short-circuit-related interruptions and demonstrate the measurable benefits of ongoing investments in fault detection, isolation, and system reinforcement across the region.

The causes of these short-circuit faults vary significantly by country, reflecting local climate and infrastructure characteristics. Environmental causes dominate in Latvia by 42% and Iceland by 64%, while lightning is the leading cause in Sweden by 31%. External influences such as third-party damage reach 40% in Lithuania, and 26% in Denmark, and technical equipment failures are the most prominent reason in Estonia 24%. This heterogeneous distribution highlights the importance of country-specific risk assessments and mitigation strategies, ranging from advanced lightning shielding and automated vegetation control to enhanced asset monitoring, in order to reduce short-circuit frequency and limit their propagation into major disturbances [3].

Before moving the overhead lines in distribution grids to underground cable, a couple of partial blackouts happened in Denmark. On 8 January 2005, hurricane-strength winds caused widespread damage to distribution lines through fallen trees and flying objects, leaving approximately 200,000 households without power. The 3 December 1999 "hurricane of the century" similarly resulted in the loss of electricity to around 440,000 households, predominantly from cable failures in the distribution grid. On 28 December 2002, two independent errors within the western Danish transmission grid interrupted supply to about one million people in northern and western Jutland for up to three hours. The most technically complex event occurred on 23 September 2003, when a double busbar fault at a South Swedish switching station tripped four 400 kV circuits and two nuclear units at Ringhals, precipitating a voltage collapse that affected the entire southern Swedish and East Danish networks [4].

These incidents illustrate both the external and internal mechanisms that initiate and propagate short-circuit-related blackouts. Weather-induced line-to-ground faults dominated the 1999 and 2005 events, underscoring the need for accurate short-circuit current calculations that account for variable fault impedances under storm conditions. In contrast, the 2003 busbar fault and the 2002 transmission-grid maloperations demonstrate how protection miscoordination or equipment failure can rapidly escalate a local symmetrical or asymmetrical short circuit into a system-wide collapse.

5.6.1 Symmetrical Faults

The symmetrical faults are often known as balanced faults. In this case, all three phase lines are connected to each other or ground equally, and the system remains in a balanced condition. These types of faults are rare in the power system, and it contributes 2–5% of the total fault. The symmetrical faults are classified as three line-to-ground (LLLG) and three-line (LLL) faults.

From transient point of view, initially, after the fault moment, the magnitude of the short circuit current will rise, as shown in Fig. 5.5. Then, the magnitude of this current will reduce to steady state value. This short circuit current is divided into three periods, namely sub transient, transient, and steady-state periods.

- **Sub transient period** (first few cycles): The current decays very rapidly. The rms value of the AC component during this period is called the **subtransient current** I'', produced mainly by the damper windings. The corresponding reactance is the **subtransient reactance** X_d''.
- **Transient period** (next several cycles): The current decays more slowly. The rms value of the AC component is the **transient current** I', associated with the transient component of the field current. The corresponding reactance is the **transient reactance** X_d'.

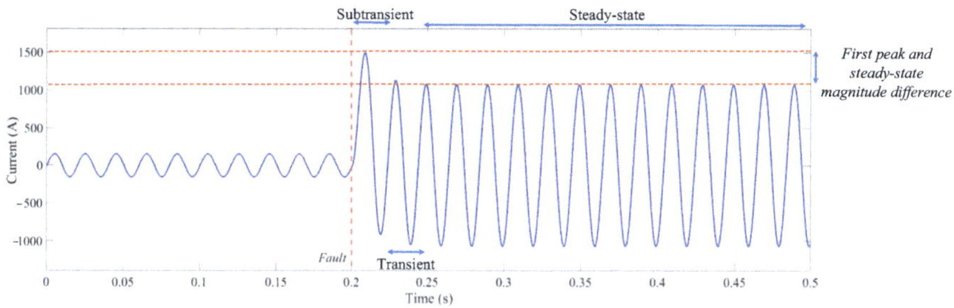

Fig. 5.5 Transient waveform of fault current

- **Steady-state period**: The current achieves a stable sinusoidal value. The rms value is the **steady-state current** I, limited by the **synchronous (direct-axis) reactance** X_d.

The subtransient, transient and steady-state currents are related to the internal generated voltage E_g (pre-fault voltage) by [5]:

$$I'' = \frac{E_g}{X_d''}, I' = \frac{E_g}{X_d'}, I = \frac{E_g}{X_d} \tag{5.23}$$

The instantaneous rms value of the AC fault current as a function of time can be expressed as:

$$I(t) = \left(I'' - I'\right)e^{-t/T''} + \left(I' - I\right)e^{-t/T'} + I \tag{5.24}$$

where T'' is the subtransient time constant and T' is the transient time constant. Immediately, after the fault, $t = 0$, the value of the current will be equal to I'', and after infinite time, the value of the current will be equal to I.

At the instant of fault inception, a DC offset component also appears, making the asymmetrical peak current approximately 1.5–2.0 times the subtransient rms value, depending on the point-on-wave of fault initiation. Figure 5.5 shows the transient waveform during faults [6].

Example 1 A three-phase 100 MVA, 11 kV, 50 Hz synchronous generator experiences a bolted three-phase fault at its terminals. The per-unit reactances are $X_d'' = 0.1$ pu, $X_d' = 0.2$ pu and $X_d = 1$ pu. The subtransient time constant $T'' = 0.05$ s and transient time constant $T' = 1.04$ s. The initial DC component is assumed to be 50% of the initial AC component.

Base current:

$$I_b = \frac{100 \times 10^6}{\sqrt{3} \times 11 \times 10^3} = 5248.63\ \text{A}$$

Subtransient current:

$$I'' = \frac{1}{0.1} = 10\,\text{pu} \Rightarrow I'' = 10 \times 5248.63 = 52486.3\,\text{A}$$

Transient current:

$$I' = \frac{1}{0.2} = 5.0\,\text{pu} \Rightarrow I' = 5.0 \times 5248.63 = 26243\,\text{A}$$

Steady-state current:

$$I = \frac{1}{1} = 1\,\text{pu} \Rightarrow I = 1 \times 5248.63 = 5248.6\,\text{A}$$

Initial peak (total) current (including DC offset):

$$I_{\text{peak}} = 1.5 \times I'' = 78729\,\text{A}$$

AC component as a function of time:

$$I(t) = 26243\,e^{-t/0.05} + 20995\,e^{-t/1.04} + 5248\,\text{A}$$

For calculations of fault current in steady-state, we can use:

$$I_F(\text{pu}) = \frac{V_m}{X_m}$$

where V_m is the pre-fault voltage in pu (usually 1.0 pu) and X_m is the total equivalent reactance to the fault point. Short-circuit kVA at bus m:

$$\text{kVA}_{sc} = \frac{V_m \times V_{Lm} \times S_b}{X_m \times V_b} = \frac{S_b}{X_m}(\text{when } V_m = V_b = 1.0\,\text{pu})$$

In percentage terms:

$$\text{kVA}_{sc} = \text{Base kVA} \times \frac{100}{\%\,X_m}$$

Example 2 A 10 MVA transformer has 10% reactance. Calculate short-circuit MVA on the LV side.

$$\text{Short - circuit MVA} = \frac{100 \times 10}{10} = 100\,\text{MVA}$$

Fault current at 11 kV:

$$I_{sc} = \frac{100}{\sqrt{3} \times 11} = 5.248\,\text{kA}$$

5.6.2 Symmetrical Components

Symmetrical components provide a powerful mathematical tool for analysing unsymmetrical faults in three-phase power systems. Any unbalanced three phase systems can be resolved into three balanced systems known as the positive-sequence, negative-sequence, and zero-sequence components. The positive-sequence components consist of three phases of equal magnitude displaced by 120° from each other and following the normal abc phase sequence. The negative-sequence components are also three equal-magnitude phases displaced by 120°, but they follow the reverse phase sequence acb. The zero-sequence components are three identical phases that are equal in both magnitude and phase angle.

To facilitate the representation of these phase shifts, the complex operator a is commonly used. This operator rotates a phasor by 120° in the anticlockwise direction and is defined as [7]:

$$
\begin{aligned}
a &= 1\angle 120^\circ = -0.5 + j0.866 \\
a^2 &= 1\angle 240^\circ = -0.5 - j0.866 \\
a^3 &= 1
\end{aligned}
\tag{5.25}
$$

and the important relation

$$
a + a^2 + a^3 = 0. \tag{5.26}
$$

The positive-sequence currents can be expressed as

$$
\begin{aligned}
I_{a1} &= I_{a1}\angle 0^\circ \\
I_{b1} &= a^2 I_{a1} \\
I_{c1} &= a I_{a1}
\end{aligned}
\tag{5.27}
$$

The negative-sequence currents are

$$
\begin{aligned}
I_{a2} &= I_{a2}\angle 0^\circ \\
I_{b2} &= a I_{a2} \\
I_{c2} &= a^2 I_{a2}
\end{aligned}
\tag{5.28}
$$

The zero-sequence currents are equal in all phases:

$$
I_{a0} = I_{b0} = I_{c0} \tag{5.29}
$$

These definitions apply to voltages as well.

Any unbalanced set of phase currents can be expressed as the sum of the three sequence components:

$$\begin{aligned} I_a &= I_{a0} + I_{a1} + I_{a2}, \\ I_b &= I_{b0} + I_{b1} + I_{b2}, \\ I_c &= I_{c0} + I_{c1} + I_{c2}. \end{aligned} \tag{5.30}$$

In matrix form this becomes

$$\begin{bmatrix} I_a \\ I_b \\ I_c \end{bmatrix} = \begin{bmatrix} 1 & 1 & 1 \\ 1 & a^2 & a \\ 1 & a & a^2 \end{bmatrix} \begin{bmatrix} I_{a0} \\ I_{a1} \\ I_{a2} \end{bmatrix} \tag{5.31}$$

The inverse transformation is

$$\begin{bmatrix} I_{a0} \\ I_{a1} \\ I_{a2} \end{bmatrix} = \frac{1}{3} \begin{bmatrix} 1 & 1 & 1 \\ 1 & a & a^2 \\ 1 & a^2 & a \end{bmatrix} \begin{bmatrix} I_a \\ I_b \\ I_c \end{bmatrix}, \tag{5.32}$$

which yields explicit expressions

$$\begin{aligned} I_{a0} &= \frac{1}{3}(I_a + I_b + I_c), \\ I_{a1} &= \frac{1}{3}\left(I_a + aI_b + a^2 I_c\right), \\ I_{a2} &= \frac{1}{3}\left(I_a + a^2 I_b + aI_c\right). \end{aligned} \tag{5.33}$$

The same relationships hold for voltages:

$$\begin{aligned} V_{a0} &= \frac{1}{3}(V_a + V_b + V_c), \\ V_{a1} &= \frac{1}{3}\left(V_a + aV_b + a^2 V_c\right), \\ V_{a2} &= \frac{1}{3}\left(V_a + a^2 V_b + aV_c\right). \end{aligned} \tag{5.34}$$

In a Y-connected system the neutral current is related to the zero-sequence component by

$$I_n = I_a + I_b + I_c = 3I_{a0}. \tag{5.35}$$

Example 3 A three-phase system has phase voltages $V_a = 92$ kV $\angle 5°$, $V_b = 68$ kV $\angle -92°$, and $V_c = 45$ kV $\angle 78°$. Calculate the symmetrical voltage components.

The zero-sequence voltage component is

$$V_{a0} = \frac{1}{3}\left(92\angle 5 + 68\angle - 92^{\circ} + 45\angle 78^{\circ}\right) = 33.3\angle - -9.2^{\circ}\ \text{kV}.$$

The positive-sequence voltage component is

$$V_{a1} = \frac{1}{3}(92\angle 5 + 68\angle(120 - 92) + 45\angle(240 + 78)) = 61.6\angle 3.05^{\circ}\ \text{kV}.$$

The negative-sequence voltage component is

$$V_{a2} = \frac{1}{3}(92\angle 5 + 68\angle(240 - 92) + 45\angle(120 + 78)) = 10.34\angle 106\ \text{kV}.$$

This decomposition clearly demonstrates how any unbalanced set of phase voltages can be successfully resolved into three balanced positive-, negative-, and zero-sequence components.

5.6.3 Sequence Impedance of Power System Equipment

The analysis of unsymmetrical faults using symmetrical components requires the determination of positive-sequence, negative-sequence, and zero-sequence impedances for each power system element. These sequence impedances differ significantly depending on the type of equipment and their grounding methods.

Synchronous machines, including generators and motors, exhibit distinctly different impedances in each sequence. The positive-sequence impedance is the normal operating impedance seen during balanced conditions and includes the synchronous reactance. For fault studies, subtransient reactance is typically used for initial current magnitude, while transient and synchronous values apply for longer durations. The negative-sequence impedance is generally lower than the positive-sequence value and is approximately equal to the subtransient reactance in the quadrature axis. Typical values range from 0.1 to 0.25 pu, often close to the subtransient direct-axis reactance. The zero-sequence impedance is usually much smaller than the positive- and negative-sequence impedances and depends strongly on the neutral grounding arrangement. With a solidly grounded neutral, the zero-sequence impedance is very low, typically in the range 0.05–0.20 pu, limited mainly by stator winding leakage. With an ungrounded or high-impedance grounded neutral, the zero-sequence impedance becomes very large, approaching infinity in ideal ungrounded cases, severely limiting zero-sequence current.

Transformer sequence impedances depend on the winding connection (Δ–Y, Y–Y, Δ–Δ, etc.) and grounding. For most practical purposes, the positive-sequence and negative-sequence impedances are equal.

The zero-sequence impedance varies significantly with connection type. In delta–delta or wye–wye ungrounded configurations, zero-sequence currents cannot flow through the transformer, resulting in an open circuit in the zero-sequence network. In a wye–wye solidly grounded on both sides, the zero-sequence impedance approximately equals the positive- and negative-sequence impedances. In wye–delta or delta–wye configurations, zero-sequence current can circulate in the delta winding but does not appear on the wye side unless the wye is grounded. The diagram of zero-sequence impedances for different transformer connection types is shown in Fig. 5.6.

Overhead transmission lines and cables have sequence impedances that differ due to conductor geometry and ground return paths. The positive sequence, zero-sequence, and negative-sequence impedances are essentially equal, as shown in Fig. 5.7.

Three-phase loads are usually modeled with equal positive- and negative-sequence impedances, while zero-sequence impedance depends on whether the load is connected in a delta or wye-grounded configuration, as shown in Fig. 5.8. Balanced three-phase loads have equal positive-, negative-, and zero-sequence impedances in delta-connected or ungrounded wye arrangements. For the zero-sequence model, due to the grounding impedance, three times the ground impedance should be connected in series with the load impedances.

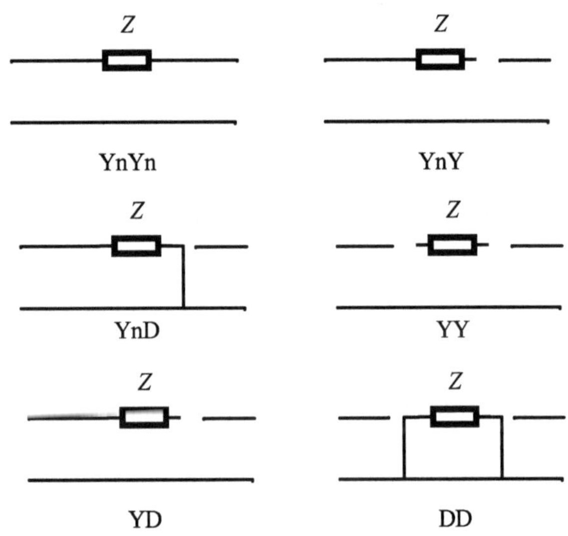

Fig. 5.6 Zero sequence model of transformers

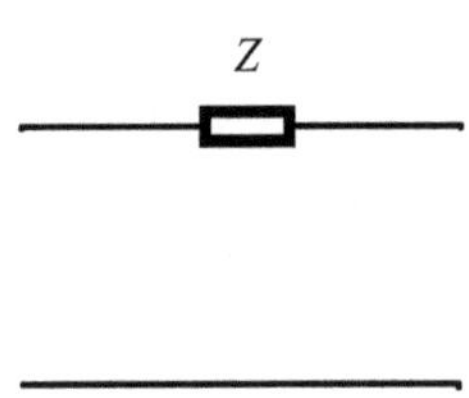

Fig. 5.7 Zero, positive, and negative sequence models for transmission lines

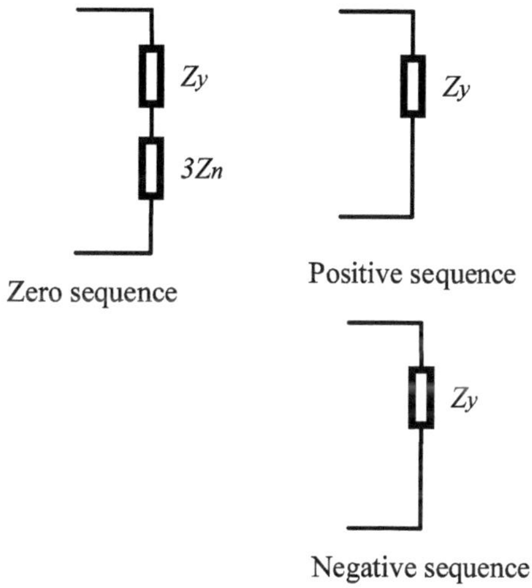

Fig. 5.8 Zero, positive, and negative sequence models for loads

5.6.4 Unsymmetrical Faults

Unsymmetrical faults in power systems are categorized as single-line-to-ground (SLG), line-to-line (LL), and double-line-to-ground (DLG) faults. These faults are analyzed using the method of symmetrical components, which resolves unbalanced voltages and currents into positive, negative, and zero sequence components.

5.6.4.1 Single-Line-to-Ground Fault

A SLG fault occurs when only one phase conductor contacts the ground, commonly through a fault impedance Z_f. For an unloaded synchronous generator with its neutral grounded solidly, as shown in Fig. 5.9, consider a fault on phase a. The boundary conditions are $V_a = 0$, $I_b = 0$, and $I_c = 0$.

Using symmetrical components, the positive sequence current is given by:

$$I_{a1} = \frac{E_a}{Z_0 + Z_1 + Z_2} \tag{5.36}$$

The current fault in phase a is:

$$I_a = 3I_{a1} = \frac{3E_a}{Z_0 + Z_1 + Z_2} \tag{5.37}$$

where E_a is the pre-fault phase voltage, and Z_0, Z_1, Z_2 are the zero, positive, and negative sequence impedances, respectively.

The sequence networks are connected in series for this fault. If a fault impedance Z_f is present, the fault current becomes:

$$I_a = \frac{3E_a}{Z_0 + Z_1 + Z_2 + 3Z_f} \tag{5.38}$$

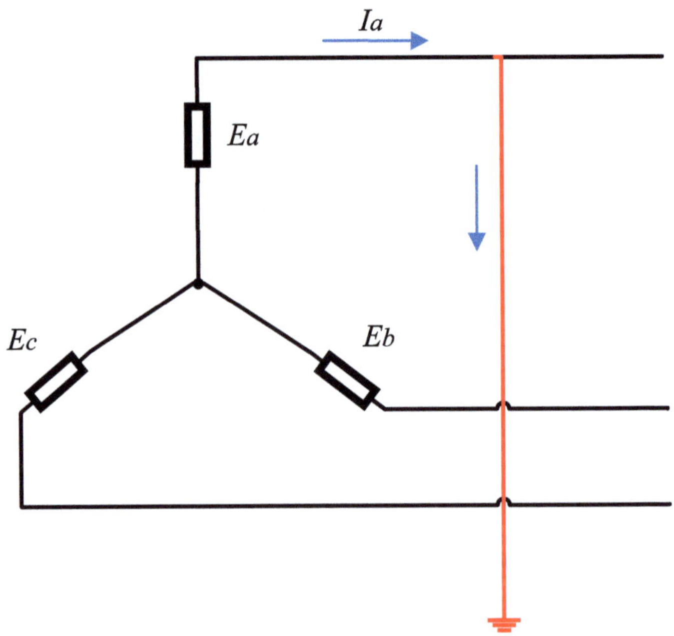

Fig. 5.9 Single-line-to-ground fault

Example 4 A three-phase 10 MVA, Y-connected, 10 kV synchronous generator is solidly grounded. The positive, negative, and zero sequence impedances are $j1.5\ \Omega$, $j0.8\ \Omega$, and $j0.3\ \Omega$, respectively. Determine the fault current in phase *a*, if an SLG fault occurs in that phase.

Solution The line-to-neutral voltage is $E_a = 10/\sqrt{3} \approx 5.77$ kV. The total impedance is $j(1.5 + 0.8 + 0.3) = j2.6\,\Omega$. Thus,

$$I_a = \frac{3 \times 5.77 \times 10^3}{2.6} \angle -90° \approx 6657\angle -90°\ \text{A}.$$

(Assuming per-phase calculation; detailed phasor computation may vary based on exact values).

5.6.4.2 Line-to-Line Fault

A LL fault occurs between two phases, say *b* and *c*, in an unloaded synchronous generator, as shown in Fig. 5.10. The boundary conditions are $I_a = 0$, $I_b = -I_c$, and $V_b = V_c$.

The sequence currents satisfy:

$$I_{a1} = -I_{a2} = \frac{E}{Z_1 + Z_2}, \tag{5.39}$$

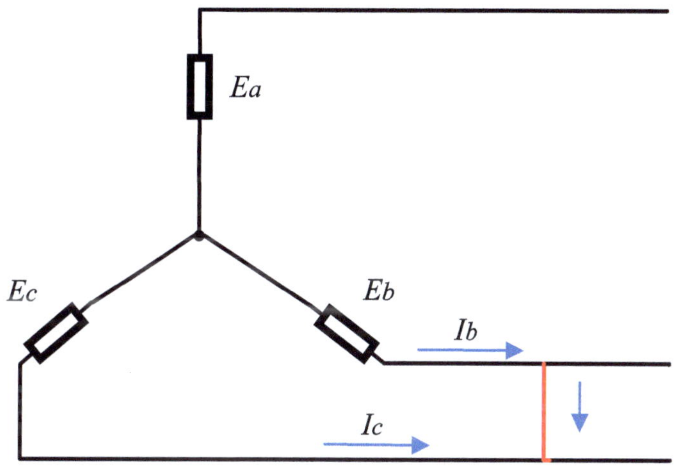

Fig. 5.10 Line-to-line fault

with $I_{a0} = 0$ (no ground involvement). The sequence networks are connected in parallel.

Example 5 The positive, negative, and zero sequence reactances of a 10 MVA, 10 kV three-phase Y-connected synchronous generator are 0.4 pu, 0.3 pu, and 0.1 pu, respectively. The neutral is solidly grounded, and the generator supplies no load. Calculate the fault current and line-to-line voltages for an LL fault between phases b and c.

Solution Base values: 10 MVA, 10 kV. Per-unit voltage $E = 1\angle 0^\circ$ pu. $I_{a0} = 0$,

$$I_{a1} = -I_{a2} = \frac{1}{j0.4 + j0.3} = -j1.42\,\text{pu}.$$

Fault current:

$$I_b = a^2 I_{a1} + a I_{a2} = 1.42\angle 150^\circ + 1.42\angle 210^\circ = -2.44\,\text{pu}.$$

Base current $I_{\text{base}} = 10 \times \frac{10^6}{\sqrt{3} \times 10 \times 10^3} \approx 577$ A, so $I_b = 1407$ A (magnitude).
Sequence voltages:

$$V_{a0} = 0,\ V_{a1} = 1 - (-j1.42)(j0.4) = 0.432\,\text{pu},\ V_{a2} = -(-j1.42)(j0.3) = 0.426\,\text{pu}.$$

Phase voltages:

$$V_a = V_{a0} + V_{a1} + V_{a2} \approx 0.858\angle 0^\circ \text{ pu},$$
$$V_b = a^2 V_{a1} + aV_{a2} \approx 0.429\angle -120^\circ + 0.426\angle 60^\circ \approx 0 \text{ pu},$$
$$V_c = aV_{a1} + a^2 V_{a2} \approx 0.429\angle 120^\circ + 0.426\angle -60^\circ \approx 0 \text{ pu}.$$

5.6.4.3 Double-Line-to-Ground Fault

A DLG fault involves two phases faulted to ground, e.g., phases b and c, as shown in Fig. 5.11. Boundary conditions: $I_a = 0$, $V_b = V_c = 0$. The sequence networks connect with positive sequences in series with the parallel of negative and zero sequences. The currents are:

$$I_{a1} = \frac{E}{Z_1 + \frac{Z_0 Z_2}{Z_0 + Z_2}},$$
$$I_{a2} = -I_{a1} \frac{Z_0}{Z_2 + Z_0}, \tag{5.40}$$
$$I_{a0} = -I_{a1} \frac{Z_2}{Z_2 + Z_0}.$$

Example 6 The positive, negative, and zero sequence reactances of a 10 MVA, 10 kV three-phase Y-connected synchronous generator are 0.20 pu, 0.19 pu, and 0.18

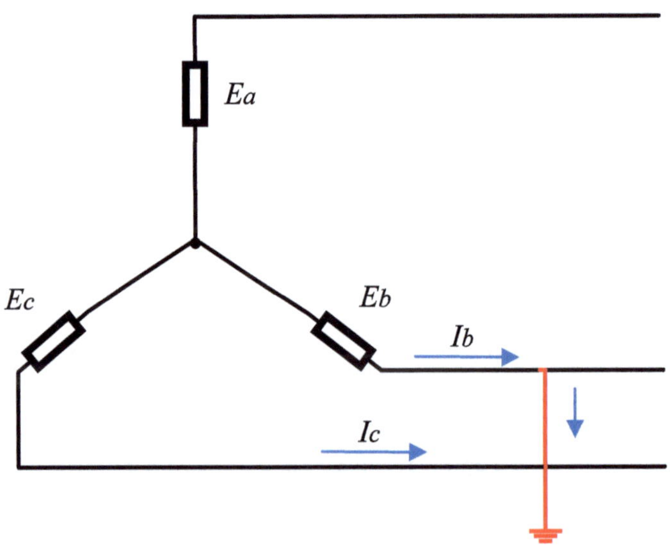

Fig. 5.11 Double-line-to-ground fault

pu, respectively. The neutral is solidly grounded. A DLG fault occurs between phases b and c. Find the subtransient currents in each phase.

Solution Base values: 10 MVA, 10 kV. $E = 1$ pu.

$$
\begin{aligned}
I_{a1} &= \frac{1}{j0.20 + \frac{(j0.18)(j0.19)}{j(0.18+0.19)}} = -j3.41\text{ pu},\\
I_{a2} &= -(-j3.41)\frac{j0.18}{j(0.19+0.18)} = j1.65\text{ pu},\\
I_{a0} &= -(-j3.41)\frac{j0.19}{j(0.19+0.18)} = j1.75\text{ pu}.
\end{aligned}
$$

Base current $I_{\text{base}} = 10 \times \frac{10^6}{\sqrt{3}\times 10\times 10^3} \approx 577$ A.
Phase currents:

$$
\begin{aligned}
I_a &= I_{a0} + I_{a1} + I_{a2} = j1.75 - j3.41 + j1.65 = 0\text{ pu},\\
I_b &= I_{a0} + a^2 I_{a1} + aI_{a2} \approx 5.11\angle 149.01^\circ\text{ pu} = 2948\text{ A}\\
I_c &= I_{a0} + aI_{a1} + a^2 I_{a2} \approx 5.11\angle 30.99^\circ\text{ pu} = 2948\text{ A}
\end{aligned}
$$

5.7 PowerFactory Analysis

In PowerFactory, the short circuit and power flow can be calculated easily by already made toolboxes. In the first window of PowerFactory, you can select different examples as shown in Fig. 5.12. In this chapter, the IEEE 14-bus system is used to apply different analyses on it. The IEEE 14-Bus System is a widely used small-scale test case. It consists of 14 buses (nodes), 5 generators (2 synchronous machines and 3 synchronous compensators), 11 loads, and 16 transmission lines, as shown in Fig. 5.13.

5.7.1 Power Flow Calculations

The load flow can be calculated by using the calculated load flow toolbox in the top left side of the toolbar, as shown in Fig. 5.14.

Then, the load flow calculation window will be opened, as shown in Fig. 5.15. In this window, three different calculation methods can be selected: AC load flow for balanced systems, unbalanced systems, and DC load flow. DC load flow will provide faster results, but simpler outcomes. Also, the automatic tap changers can be checked in this window to allow tap changers to automatically adjust the voltage.

By selecting the calculation settings, as shown in Fig. 5.16, the number of iterations for the Newton-Raphson method can be selected. Higher iterations will lead

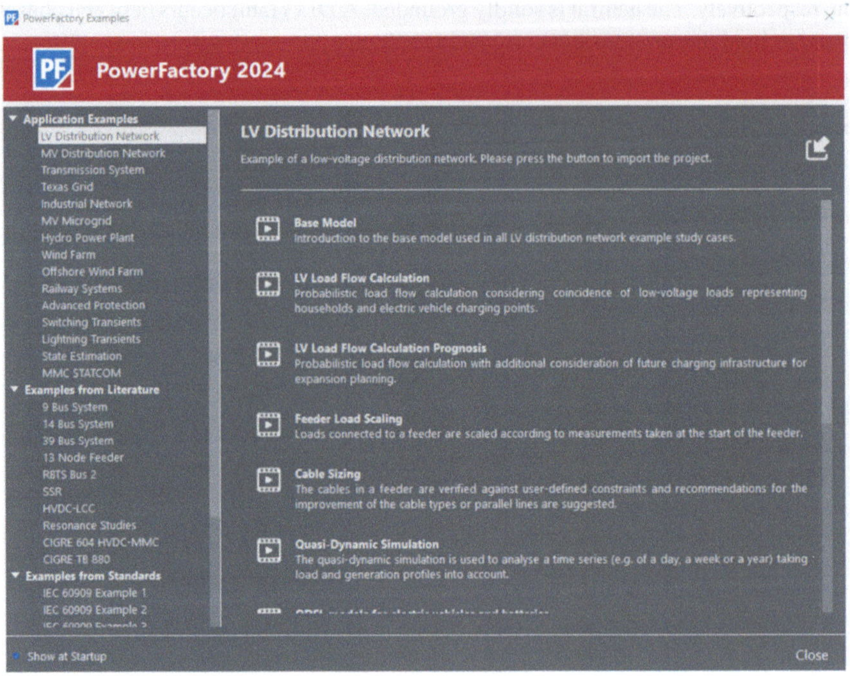

Fig. 5.12 PowerFactory example window

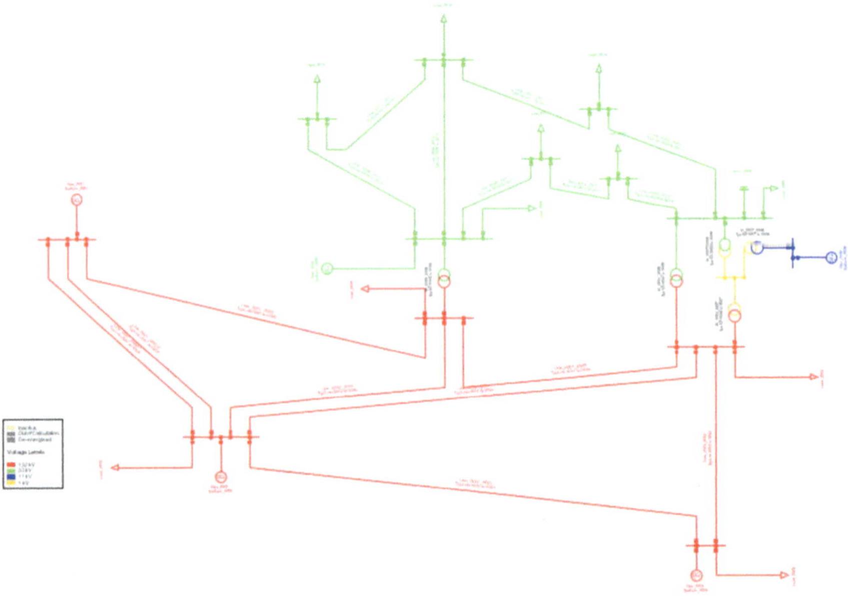

Fig. 5.13 IEEE 14-bus system

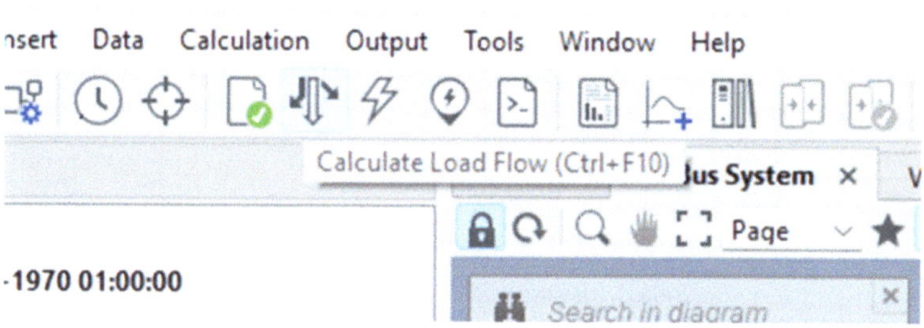

Fig. 5.14 Calculate load flow

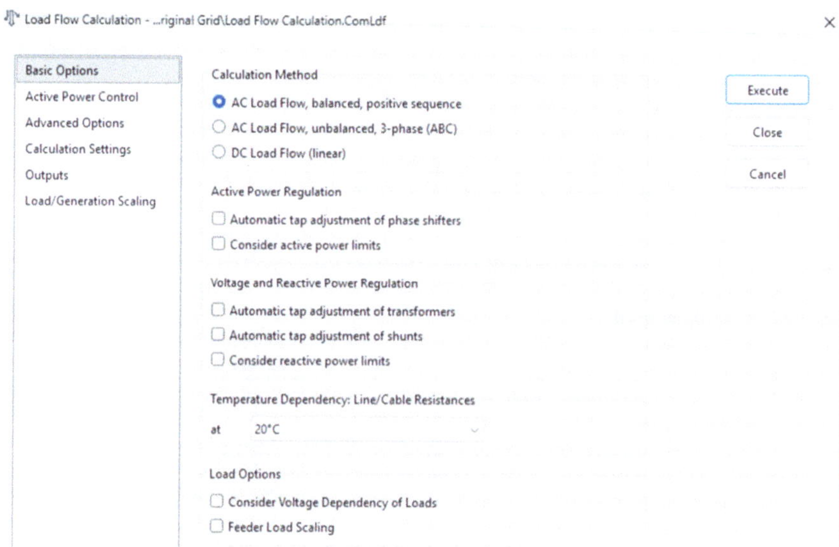

Fig. 5.15 Load flow calculation window

to more accurate results; however, it will make the calculation slower. By selecting the execute, the results of active, and reactive powers, voltages, and currents will be shown in an edge box at each component in the grid window, as shown in Fig. 5.17.

5.7.2 Steady-State Short-Circuit Analysis

The short-circuit analysis can be done by right-clicking on any equipment and selecting calculation and short-circuit, as shown in Fig. 5.18. It will provide the steady state results for a short-circuit event. In the new window, as shown in

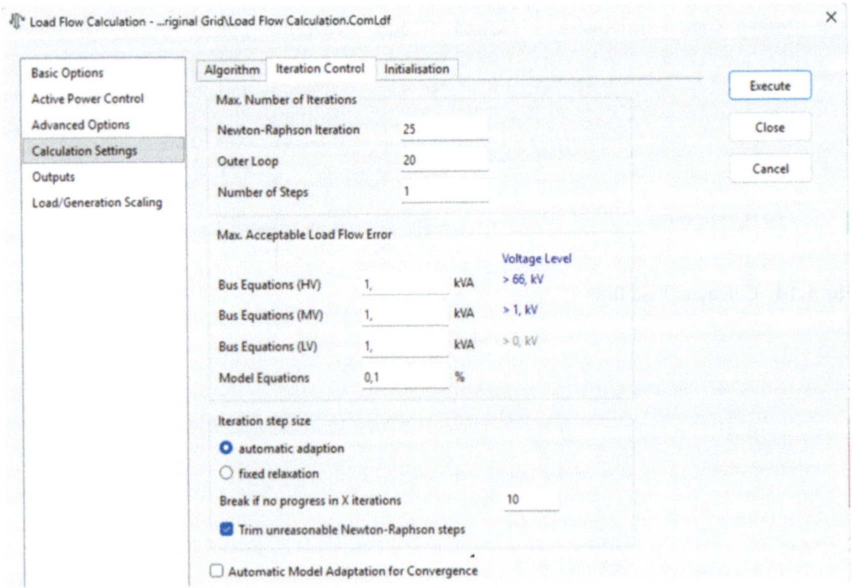

Fig. 5.16 Load flow calculation settings

Fig. 5.17 Load flow results

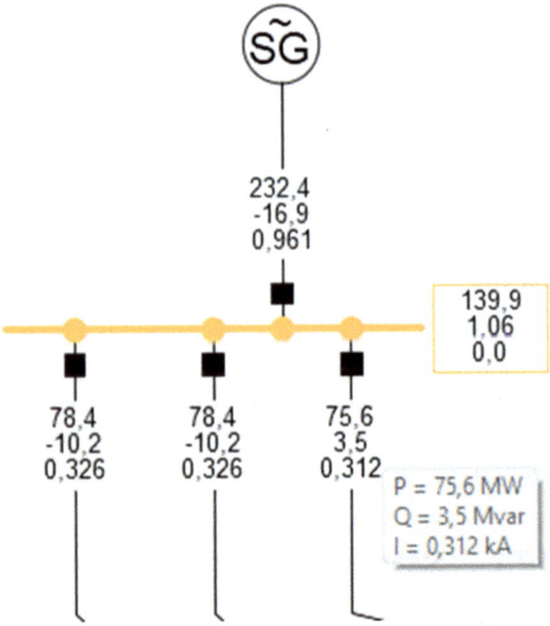

Fig. 5.19, the fault resistance and location of the short-circuit in the line, for example, 50% of the line, can be selected.

Also, as shown in Fig. 5.20, the fault type can be selected among different types. Note that PowerFactory cannot simulate the fault from phase to neutral if there is no predefined neutral. After executing the calculations, results can be shown in the grid's window, as shown in Fig. 5.21. It will provide the steady-state current, in addition to short-circuit MVA.

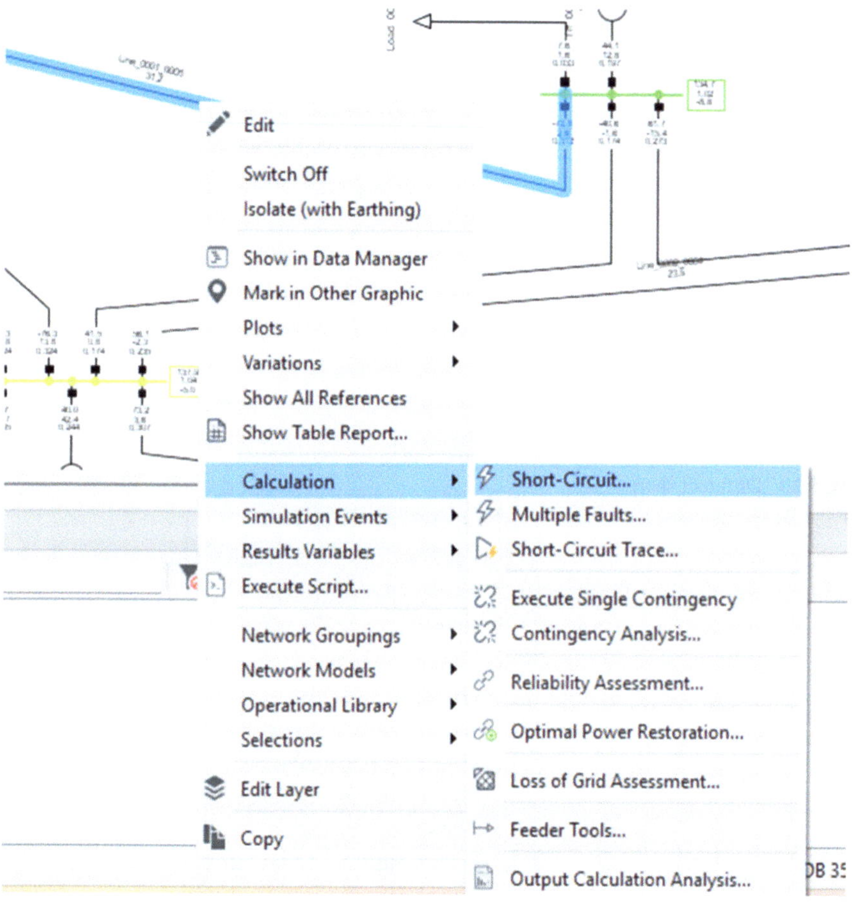

Fig. 5.18 Short circuit calculation

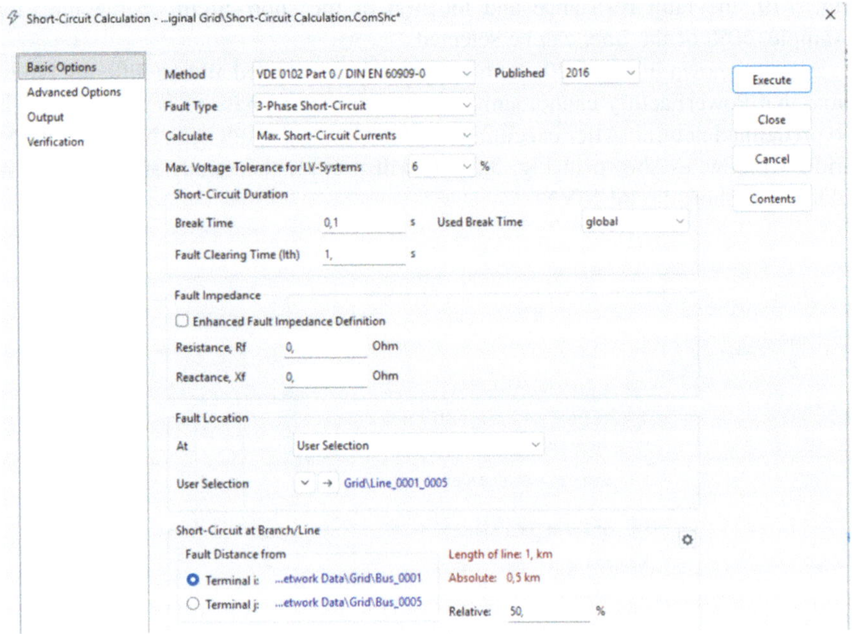

Fig. 5.19 Short circuit calculation setting

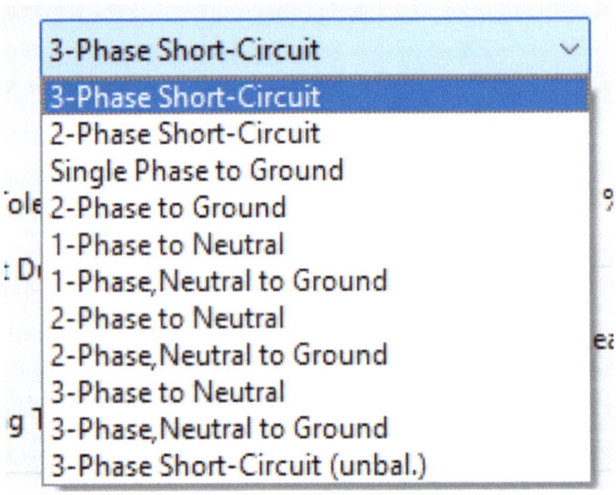

Fig. 5.20 Different short circuit types

5.7.3 Transient EMT/RMS Simulation

To run the EMT/RMS simulations for transient studies, a Simulation RMS/EMT toolbox is defined in PowerFactory. First, an event should be defined. By right-clicking on one component, for example, as shown in Fig. 5.22, a transmission line, and selecting simulation events, short-circuit event, the specifications of a

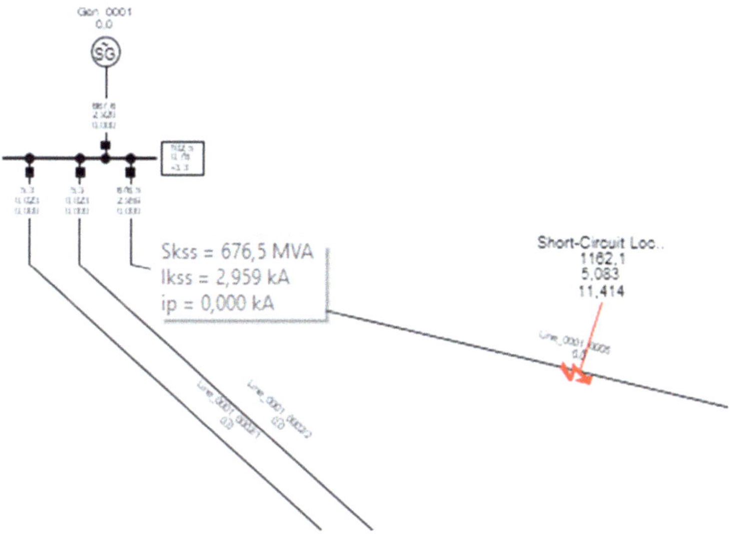

Fig. 5.21 Short circuit calculation results

fault can be defined as shown in Fig. 5.23. First, the time should be set; the fault location, type, and resistance can also be defined in this window.

After defining the event, by selecting the Simulation RMS/EMT toolbox, as shown in Fig. 5.24, and selecting the calculations of initial conditions, PowerFactory starts to make a load flow to find the initial status of the grid.

Then, in the calculation of initial conditions window, as shown in Fig. 5.25, the simulation method can be selected between RMS values, instantaneous values, or electromagnetic transients (EMT). During EMT simulations, more detailed transients of the grid can be observed. The primary difference between RMS and EMT simulations is their time resolution and detail. RMS simulation averages waveforms over a cycle, and EMT simulation solves instantaneous voltage/current waveforms to capture harmonics, switching, and fast dynamics.

After executing the initial conditions, by clicking on run simulation, the window shown in Fig. 5.26 will be opened. In this window, the duration of the simulation can be defined.

After running the simulation, by right-clicking on the line and selecting results variables, simulation RMS/EMT as shown in Fig. 5.27, the variable of phase current for EMT simulation can be selected as shown in Fig. 5.28. Therefore, by plotting this variable, the EMT simulation waveform can be depicted as shown in Fig. 5.29. As can be seen, after 0.2 s, the current flow from the line is increased, and it will decrease to a steady-state value after passing the transients.

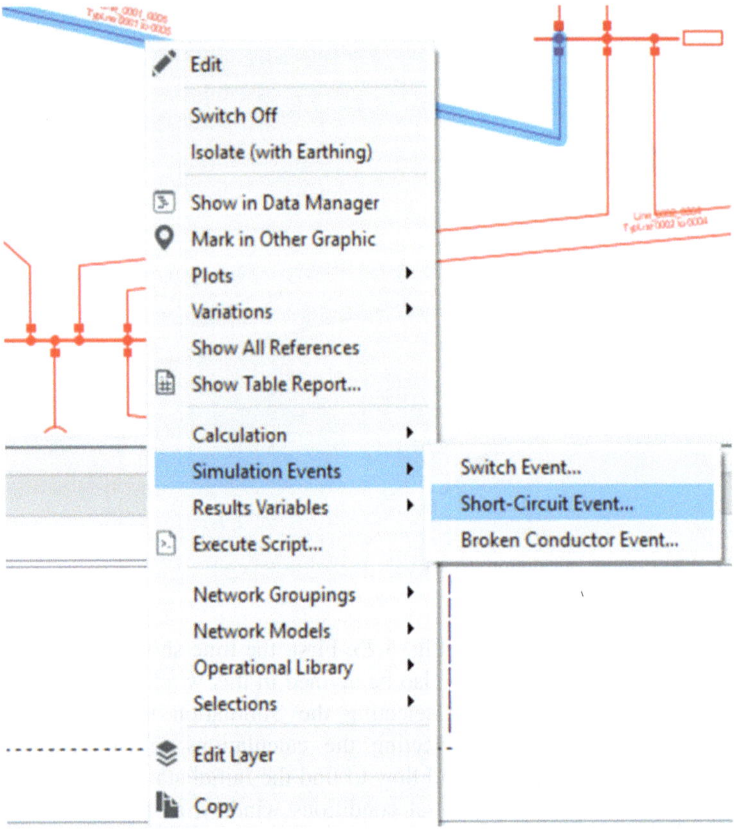

Fig. 5.22 Short circuit event

5.8 Summary

This chapter has provided a complete theoretical and practical treatment of power-flow and short-circuit calculations, the two core computational tools used for steady-state and fault analysis in modern power systems. It began by formulating the nonlinear power-balance equations, classifying buses as slack, PV, and PQ, and deriving the solution algorithms (Gauss-Seidel, Newton-Raphson, and fast decoupled methods) together with convergence criteria and practical considerations for large-scale networks. The chapter then developed symmetrical-component theory for short-circuit studies, deriving fault-current expressions for the four major fault types (three-phase, single-line-to-ground, line-to-line, and double-line-to-ground) and explaining their application in equipment rating, protection settings, and system design. Finally, detailed step-by-step tutorials in PowerFactory—DIgSILENT demonstrated how to execute balanced and unbalanced AC load-flow studies, steady-state short-circuit calculations with user-defined fault

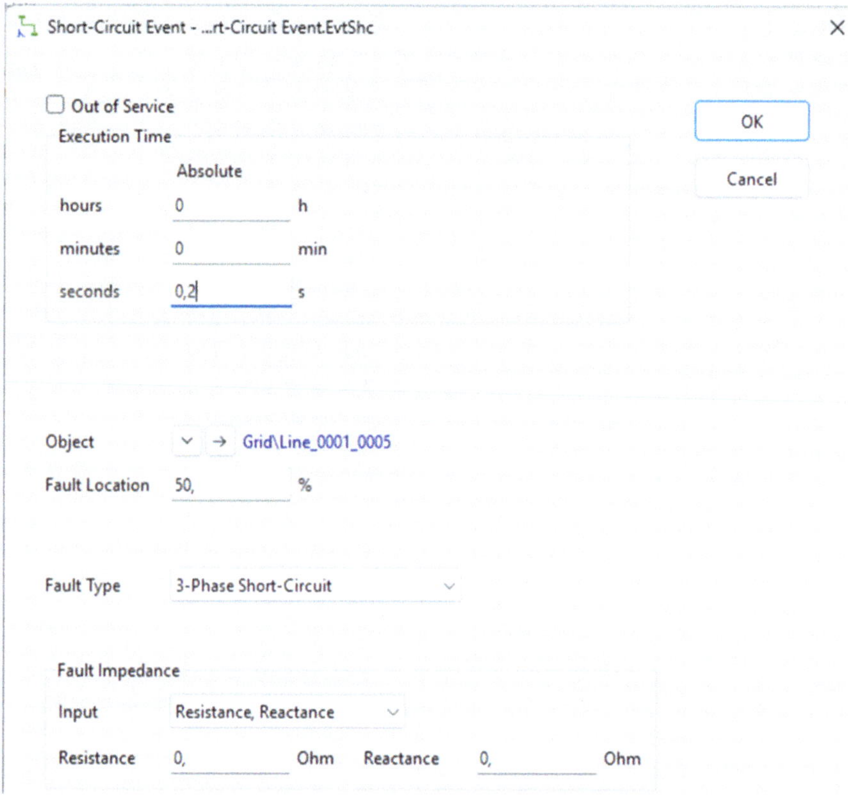

Fig. 5.23 Short circuit event window

locations and resistances, and transient RMS/EMT simulations using predefined events, result variables, and waveform plotting. The analytical methods and software skills presented in this chapter form the essential bridge to the protection and stability studies covered in Chaps. 6 and 7.

5.9 Problems and Questions

1. A simple two-bus power system operates on a 100 MVA base. Bus 1 is the slack bus with voltage $V_1 = 1.05\angle 0°$ pu. Bus 2 is a PQ bus with scheduled load $P_2 = -1.8$ pu and $Q_2 = -0.6$ pu. The line impedance between the buses is $Z_{12} = 0.05 + j0.25$ pu. Using the Gauss-Seidel iterative method and an initial guess of $V_2^{(0)} = 1.0\angle 0°$ pu, perform **one full iteration** and calculate the updated voltage $V_2^{(1)}$ at Bus 2. Show all intermediate steps, including the calculation of the injected current and the voltage correction.

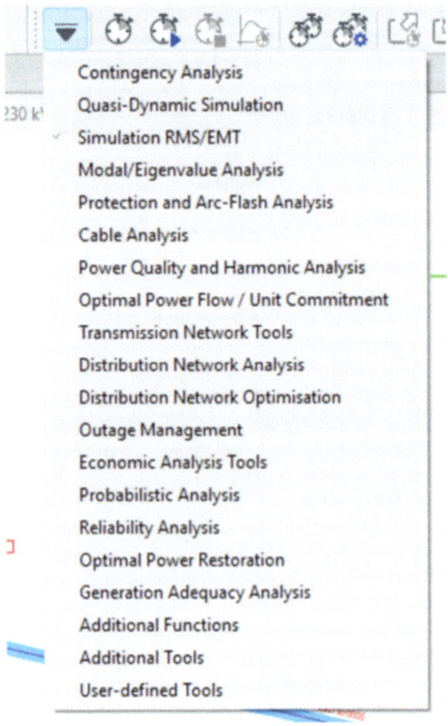

Fig. 5.24 RMS/EMT toolbox

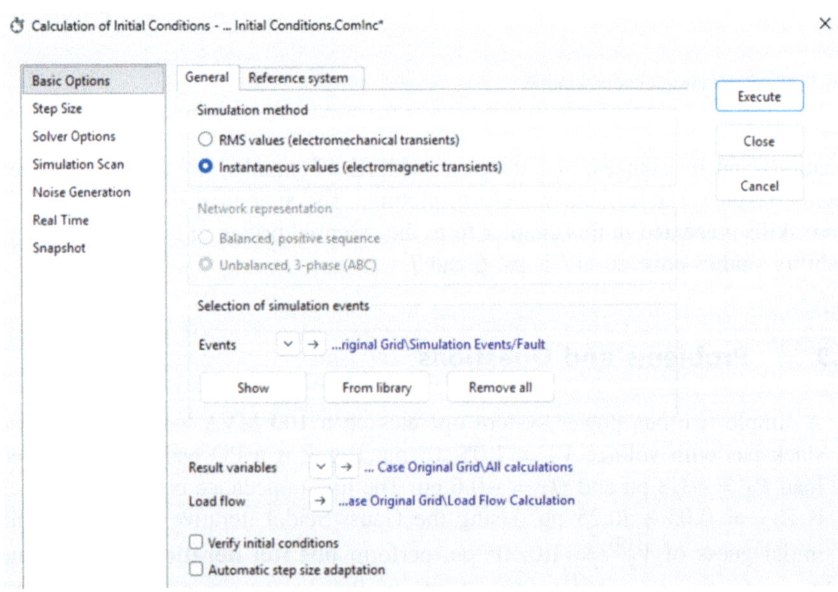

Fig. 5.25 Initial condition

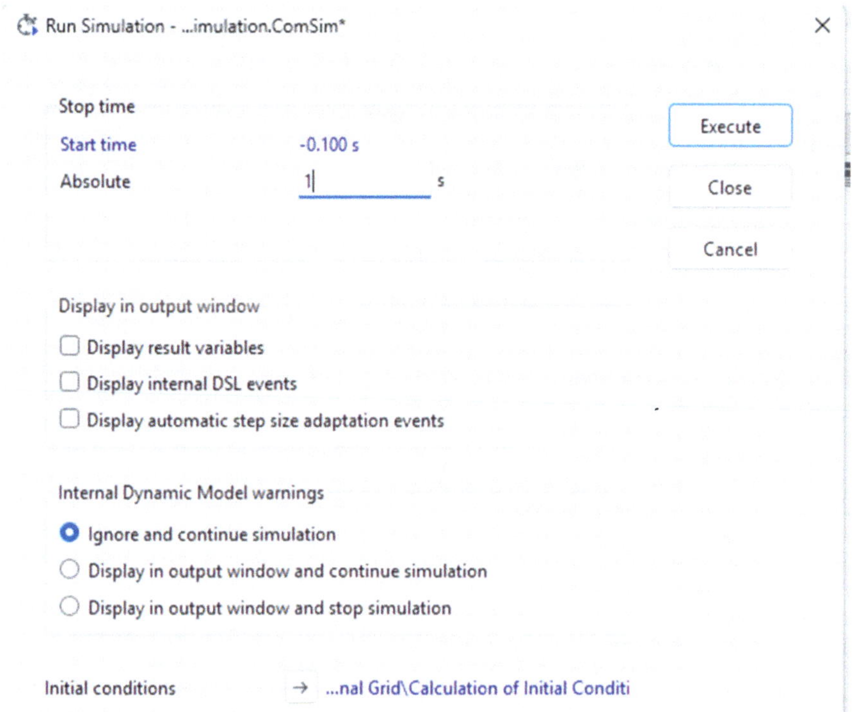

Fig. 5.26 Run simulation

2. A synchronous generator with subtransient reactance $X_d^{''} = 0.18$ pu is connected to a 132 kV bus. A single line-to-ground fault occurs at this bus. The positive-, negative-, and zero-sequence reactances of the system (including the generator) are $X_1 = 0.15$ pu, $X_2 = 0.18$ pu, and $X_0 = 0.08$ pu, respectively. Assuming the pre-fault voltage at the fault point is 1.0 pu and neglecting resistance, calculate (a) the subtransient fault current in per-unit and (b) its value in kA (use base values of 100 MVA and 132 kV line-to-line).

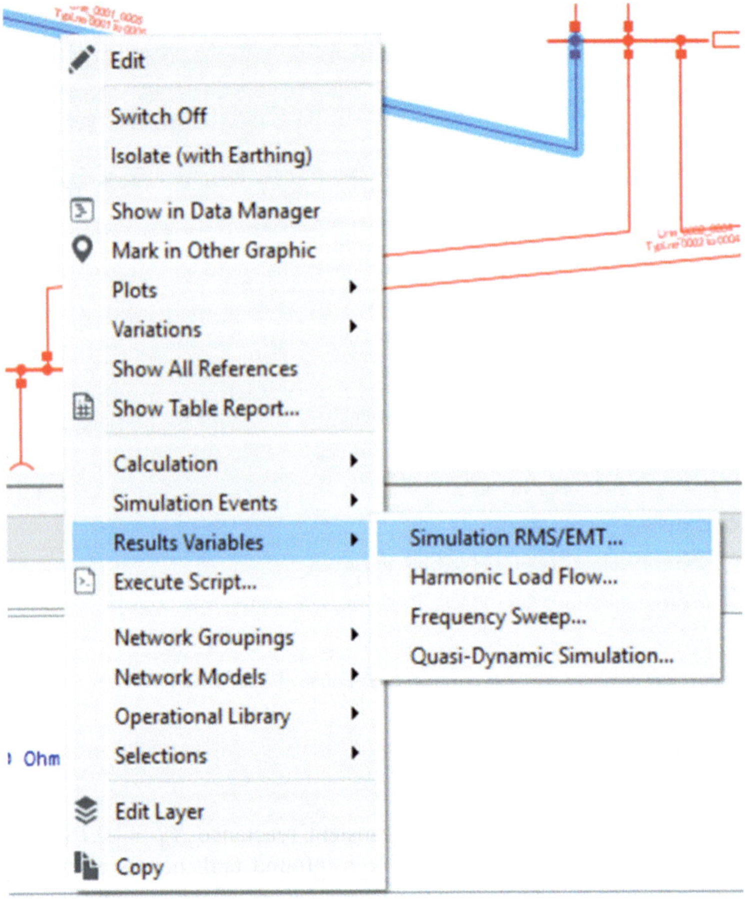

Fig. 5.27 Results variables

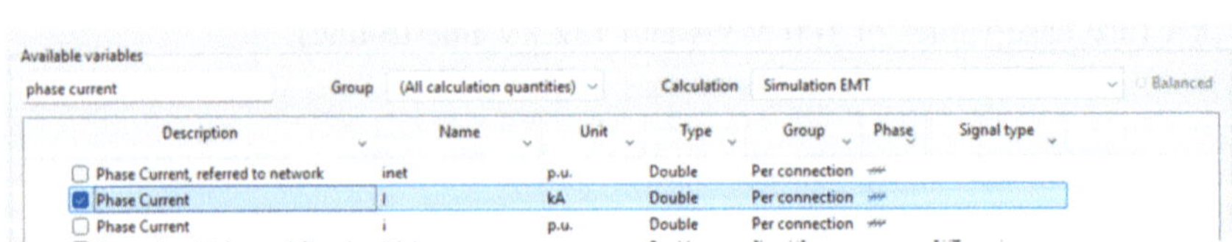

Fig. 5.28 Add phase current as result variables

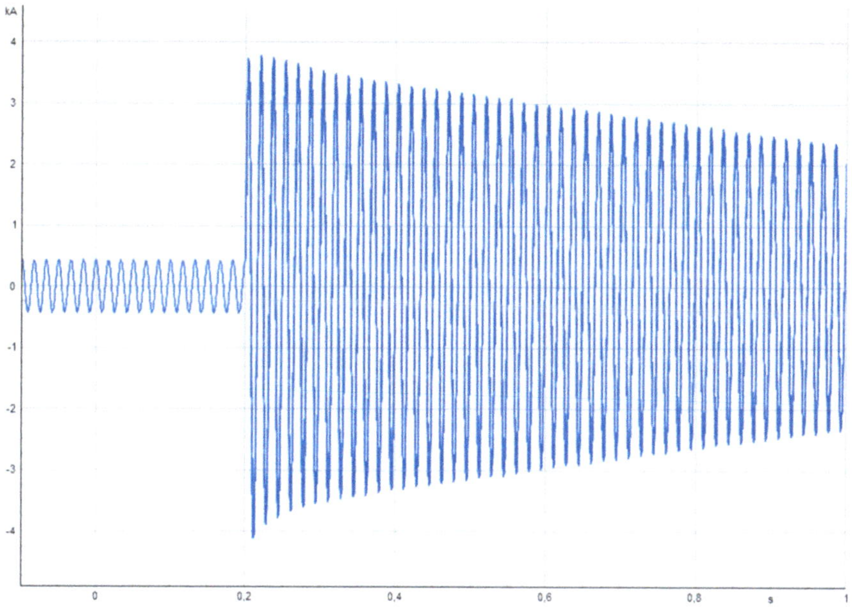

Fig. 5.29 EMT simulation for short circuit

References

1. Babu S, Shayesteh E, Hilber P (2016) Analysing correlated events in power system using fault statistics. In: 2016 international conference on probabilistic methods applied to power systems (PMAPS). IEEE, pp 1–6
2. Rozegnał B, Albrechtowicz P, Mamcarz D, Radwan-Pragłowska N, Cebula A (2020) The short-circuit protections in hybrid systems with low-power synchronous generators. Energies 14(1):160
3. 2022 Nordic and Baltic Grid Disturbance Statistics For publishing I 18 October 2023
4. Danish Energy Agency (2015) Security of electricity supply in Denmark
5. Adkins B (1951) Transient theory of synchronous generators connected to power systems. Proc IEE-Part II: Power Eng 98(64):510–523
6. Bayati N, Mortensen LK, Savaghebi M, Shaker HR (2022) A localized transient-based fault location scheme for distribution systems. Sensors 22(7):2723
7. Van der Sluis L (2001) Transients in power systems. Wiley, New York

Protection Devices

6

Abstract

This chapter provides a complete and practical introduction to the main protection devices used in power systems. It explains how these devices work together to detect faults quickly, isolate only the faulty part of the network, and protect expensive equipment such as generators, transformers, and transmission lines from serious damage. The chapter begins with fuses, describing their simple construction, fast operation under high currents, time-current behaviour, and the important rules for coordinating them in distribution networks so that only the fuse closest to the fault operates. Overcurrent relays are discussed in detail, including their different operating characteristics, how settings are selected, and the step-by-step coordination process between primary and backup relays. Differential protection is presented as one of the fastest and most selective methods, with emphasis on why bias is used and how it overcomes practical challenges like transformer inrush and measurement errors. Distance protection explained by the main operating shapes, the concept of three protection zones with different reaches and time delays, and the way engineers balance speed and backup capability. The chapter ends with clear, step-by-step guidance on implementing all these devices in PowerFactory, including fuse installation, overcurrent relay settings, and full distance relay configuration using real examples.

Keywords

Relays • Distance relay • Overcurrent relay • Fuse

N. Bayati and M. A. Khan, *Power System Fundamentals with PowerFactory*,
https://doi.org/10.1007/978-3-032-32450-4_6

6.1 Introduction

During short circuits, a high-rise current flows through lines and other equipment, which may cause melting and damage to them, and also frequency oscillation in the power system. Power system protection is a set of methods and equipment used to limit the damage caused by an electrical fault and protect other components of the grid, like generators and transmission lines, during high currents. For this purpose, a couple of components such as fuses and relays are installed in power systems.

6.2 Fuses

Fuses are among the simplest and most reliable overcurrent protective devices used in power systems. A fuse consists of a fuse element made of a low-melting-point wire that melts and interrupts the circuit during high current flows. Fuses provide fast-acting protection, particularly under short-circuit conditions, and are widely employed in distribution networks, industrial installations, and as backup protection for transformers, motors, and cables. Typically, fuses are useful for systems under 60 kV, as shown in Fig. 6.1, and due to the lower cost of fuses over other protection units, they are useful. However, fuses should be replaced after each usage, and also after changing of fault current level or system's topology, they should be replaced, due to the lack of settings.

Fuses operate in two main periods: the pre-arcing period and the arcing period. During the pre-arcing period, the fuse carries current and heats due to its resistance, absorbing electrical energy (RI^2 losses). The element eventually reaches its melting temperature, and in the arcing period, an arc forms across the melted element, and the voltage across the fuse rises sharply while the current is forced to zero. The total operation is expressed by:

$$e = iR + L\frac{di}{dt} + v_f \tag{6.1}$$

where, e is the source electromotive force, i is the instantaneous current through the fuse, R and L are the resistance and inductance of the protected circuit, respectively, and v_f is the voltage across the fuse, which is small during pre-arcing and rises sharply during arcing.

Fuses are particularly effective under high short-circuit currents. Under such conditions, only a fraction of a power–frequency cycle flows before interruption, and the current waveform includes transients influenced by circuit parameters, fault inception angle, and the current-limiting action of the fuse itself. Because prospective short-circuit currents cannot be assigned as a single value for time determination, the energy passed through the fuses can be defined by:

$$I^2t = \int_{t_0}^{t_c} i^2(t)\,dt \tag{6.2}$$

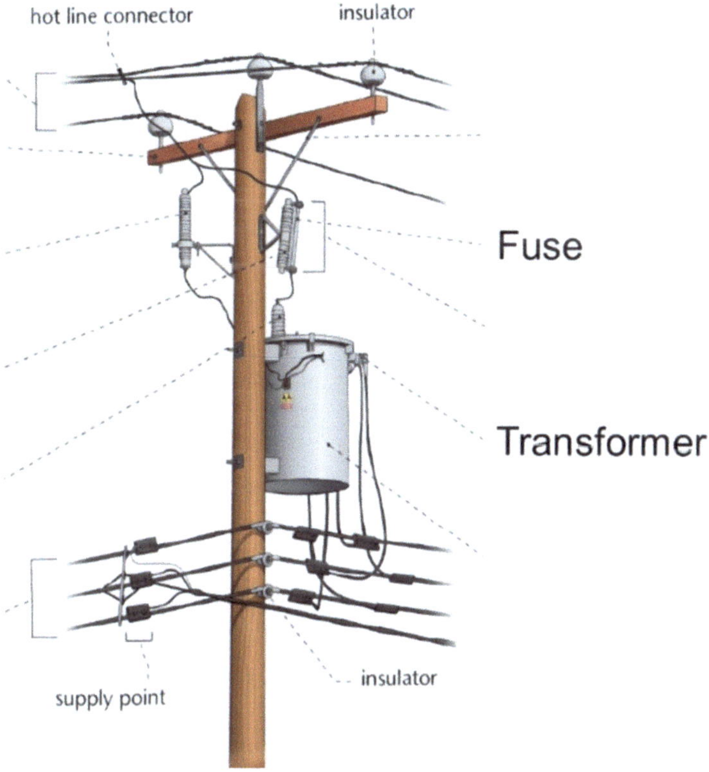

Fig. 6.1 Fuse installation in towers

This quantity represents the energy passed through the fuse from fault inception to arc extinction and is critical for assessing thermal stress and coordination.

6.2.1 Time–Current Relationships

The time–current characteristic (TCC) of a fuse shows the relationship between current magnitude and its operating time. Fuses should have TCCs that align closely with the withstanding capabilities of protected equipment, ensuring operation before damage occurs to cables, transformers, or motors, as shown in Fig. 6.2. At low overcurrents or near-rated current, fuses operate slowly to allow temporary overloads, such as motor starting. At high fault currents, operation becomes extremely rapid to provide a fast clearance.

A key parameter for high-fault-current regions is the pre-arcing I^2t, as energy to melt the element, and total clearing I^2t, as the summation of pre-arcing and arcing energy let-through. Pre-arcing I^2t depends primarily on the fuse element's

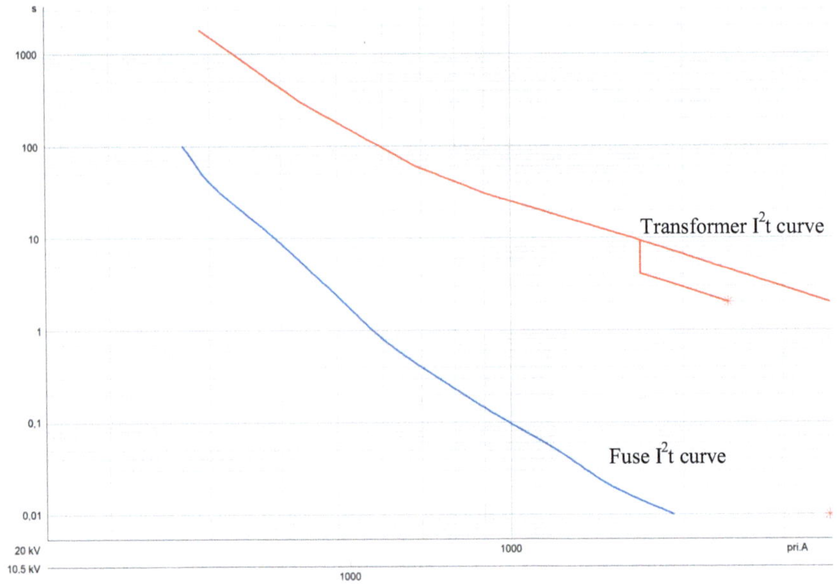

Fig. 6.2 TCC curve of fuse and transformer

cross-section and material, while total clearing I^2t also varies with system voltage, power factor, and fault level.

The TCC is typically provided by the manufacturer, including:

- Melting I^2t (pre-arcing) for short times (<10 ms)
- Clearing I^2t at specific multiples of rated current

For very high currents, fuses enter the current-limiting region, where peak current and energy are significantly reduced compared to the prospective fault current. This limits mechanical and thermal stress on downstream equipment.

Breakdown into components:

$$I^2t_{\text{total}} = I^2t_{\text{pre-arcing}} + I^2t_{\text{arcing}} \tag{6.3}$$

- **Pre-arcing I^2t**: energy required to melt the element (depends mainly on fuse geometry and material; nearly constant at very high currents)
- **Arcing I^2t**: energy during the arcing phase (depends on system voltage, power factor, and fault level)

TCCs are plotted on log–log scales, as depicted in Fig. 6.2. As can be seen, there is a downward trend, which shows fast operation time during high currents. Also, it should not operate during the rated current of the system or under overload

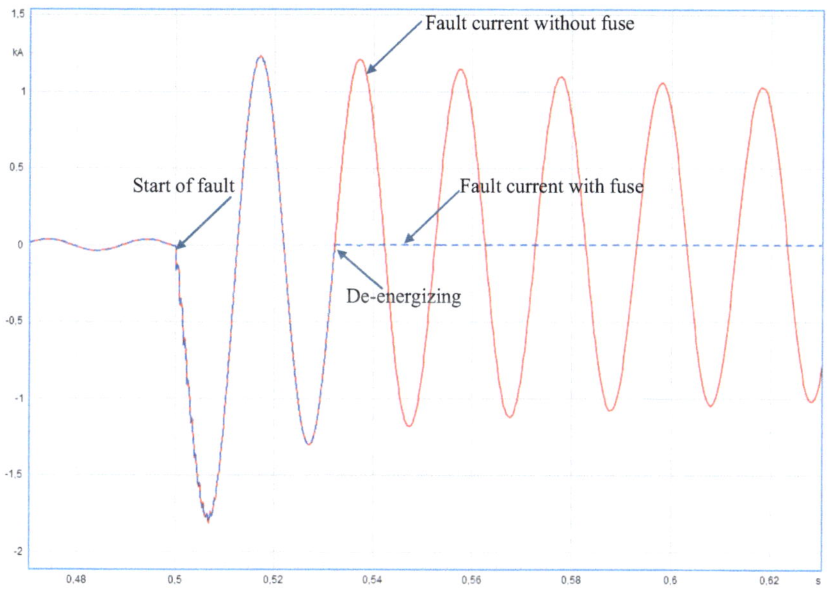

Fig. 6.3 Fault current with and without fuse operation

conditions. Typically, a 25% overloading is also considered during the selection of fuses. A fuse operation during faults is shown in Fig. 6.3. As can be seen, after a fault, it will take a couple of cycles for the operation of a fuse.

6.2.2 Fuse-Fuse Coordination

Fuse-fuse coordination is a part of a power system's selectivity to ensure that only the fuse closest to the fault operates and isolates the faulty section while keeping the rest of the system energized. Poor coordination leads to unnecessary outages of healthy parts. Therefore, in TCC, the downstream fuse should have a curve lower than the upstream fuses, or in other words:

Total clearing I^2t (downstream)<Melting (pre-arcing) I^2t (upstream)

$$I^2t_{\text{total, downstream}} < I^2t_{\text{pre-arcing, upstream}}$$

Coordination must cover overloads, low-level faults, and maximum available short-circuit currents. In cases involving transformers or motors, inrush currents and starting characteristics must also be considered to avoid nuisance operation. This ensures the downstream fuse clears the fault before the upstream fuse begins to heat significantly.

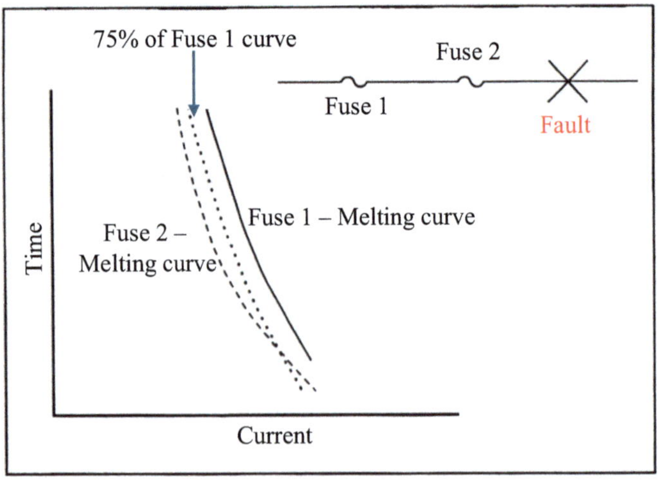

Fig. 6.4 Selectivity of fuses

To ensure the coordination between fuse, the following steps can be defined, as shown in Fig. 6.4:

1. The maximum characteristic of fuse 2 is plotted, and fuses should be selected for a case with 25% overloading. For example, if rated current is 100 A, a 125 A fuse should be selected.
2. The characteristic of fuse 1 is plotted
3. 75% of the melting curve of fuse 1 is plotted to avoid any damage to it
4. The curve of fuse 2 should be the first curve to operate.

6.2.3 Applications for Fuses

Fuses are applied across various voltage levels and system types due to their simplicity, speed, and cost-effectiveness.

- **Low-voltage distribution and industrial systems**: Protect cables, switchboards, and small loads. They offer fast short-circuit clearance and are common in panelboards and motor control centers.
- **Medium-voltage distribution**: Protect overhead lines, transformers, and capacitors. Current-limiting fuses (backup or full-range) protect pad-mounted transformers, providing high interrupting ratings and energy limitation.
- **Transformer protection**: Fuses on primary/secondary sides limit through-fault damage, typically 10 kV/0.4 kV. Backup current-limiting fuses pair with expulsion links for full-range coverage.

- **Motor circuits**: Time-delay fuses accommodate starting inrush while protecting against stalled conditions or short circuits.

For example, the Microgrid example from PowerFactory—DIgSILENT is shown in Fig. 6.5. Three fuses are installed on the lines, and the TCCs are shown in Fig. 6.6. As can be seen, the Fuse T13 has the lowest curve, as it is connected to the downstream. Thus, during a fault, Fuse T13 will operate first, and in case of maloperation, Fuse T14 will operate. For example, as can be defined from Fig. 6.6, during a 100 A fault current, Fuse T13 operates after 8 s, and Fuse 14 operates after 40 s.

6.3 Measurement Transformers

In power systems, high primary currents and voltages cannot be measured directly with standard instruments because of safety and insulation limitations. Measurement transformers step line's voltage and current down to safe secondary levels, typically 1 or 5 A for current transformers, and 110 or 220 V for potential transformers, as shown in Fig. 6.7. Then, it can be used for ammeters, voltmeters, relays, and protection devices.

6.3.1 Current Transformers

A current transformer (CT) consists of two windings: primary and secondary. The primary winding has a lower number of turns, and it is connected in series with the power line and carries the line current. The secondary winding has a higher number of turns and is connected to the ammeter or relays, as shown in Fig. 6.8. The CT operates as a step-up transformer with respect to voltage but steps down the current.

The CT is essentially a step-up transformer. Assume, N_1 is the number of turns of the primary winding, N_2 is the number of turns of the secondary winding, I_1 is primary current, and I_2 is secondary current. The basic relationship for transformers is:

$$\frac{I_1}{I_2} = \frac{N_2}{N_1} \tag{6.4}$$

Because $N_2 \gg N_1$, the secondary current is reduced to a safe value. For example, a typical ratio can be 500:5.

Example 1 A 250:5 CT is used with an ammeter. If the ammeter reads 3.3 A, the actual line current is:

$$I_1 = 3.3 \times \frac{250}{5} = 165\ \text{A}.$$

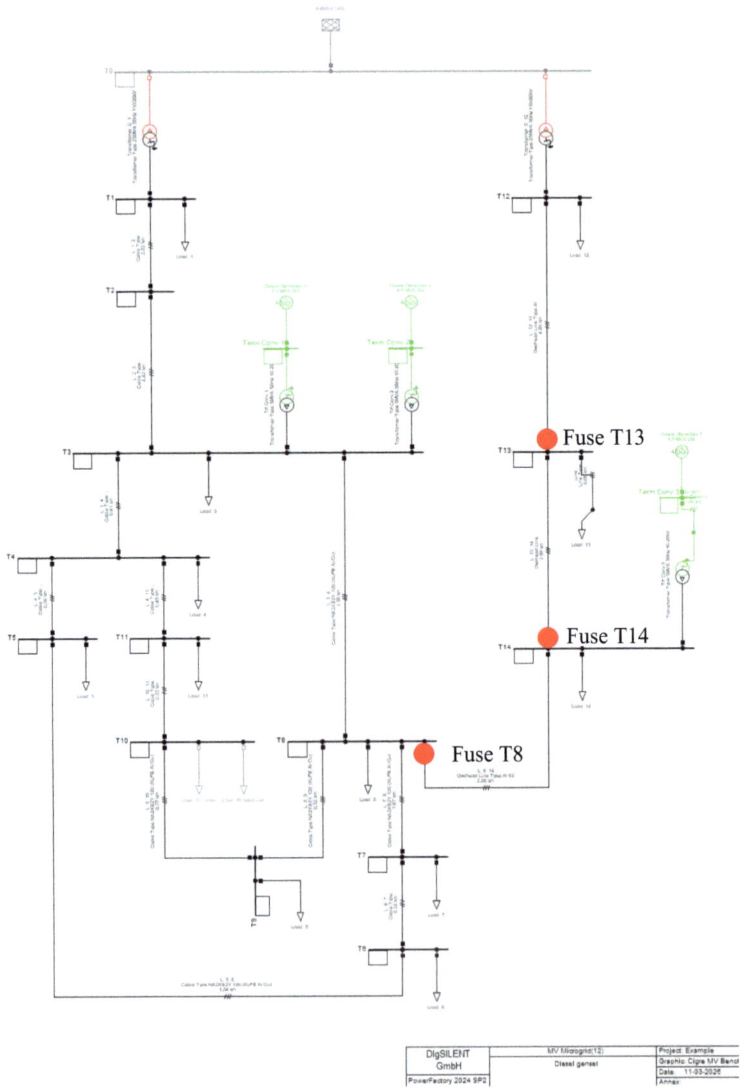

Fig. 6.5 Microgrid example from PowerFactory—DIgSILENT

The secondary of a CT must never be left open-circuited. An open secondary forces all primary ampere-turns to become magnetizing current and produce extremely high induced voltages at the secondary that can puncture insulation. This causes the CT to produce dangerously high, potentially fatal voltage spikes, saturate the core, and cause severe overheating, which poses a fire or explosion risk. These high peaks have a waveform, like a pulse, as shown in Fig. 6.9.

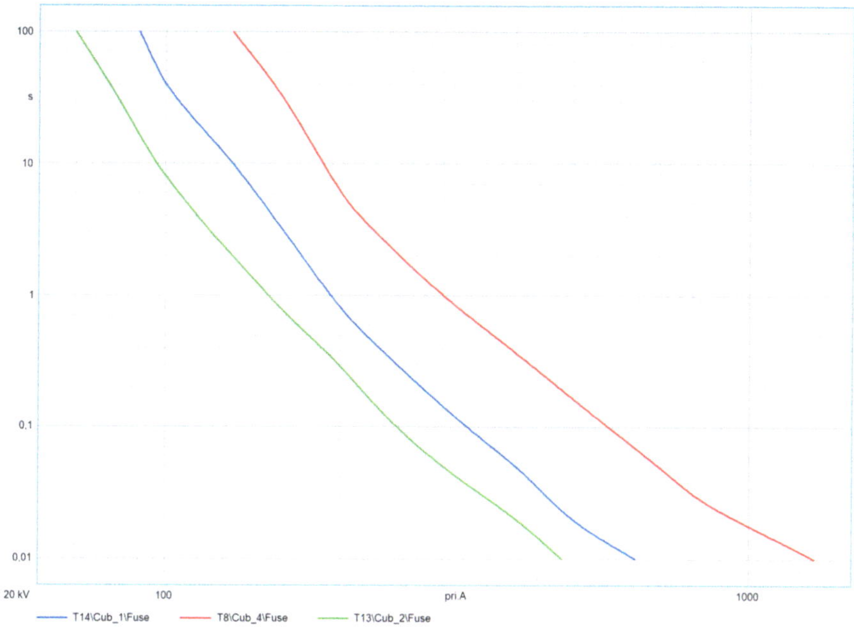

Fig. 6.6 TCCs for Microgrid example

Fig. 6.7 CT and PT connection

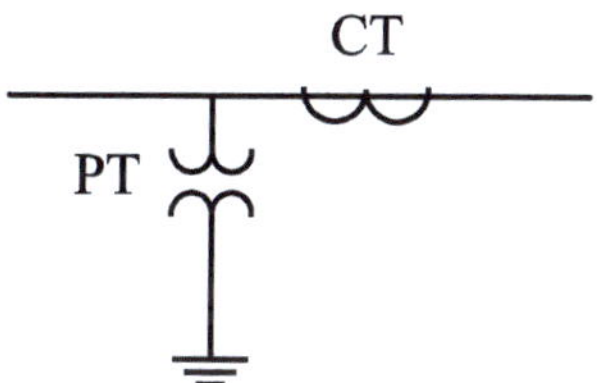

Fig. 6.8 CT connection

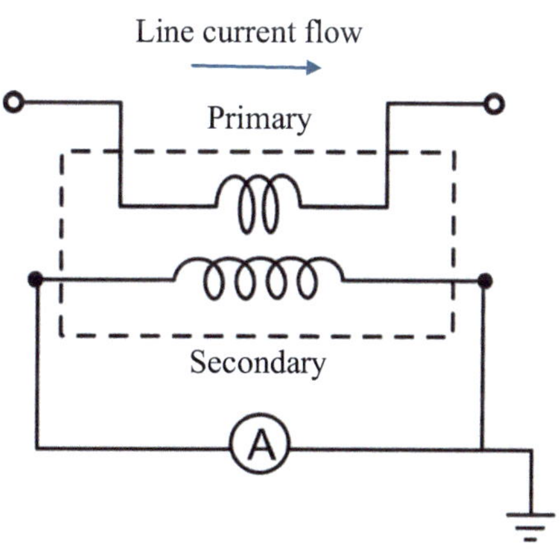

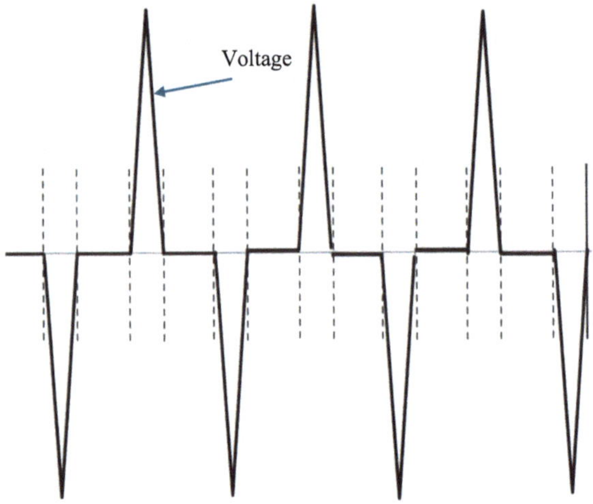

Fig. 6.9 Open circuit voltage of a CT

In theory, the transformation ratio equals the turns ratio exactly, but in practice, due to the non-ideal behavior of the transformer, it varies with load current. The percentage ratio error is:

$$\% \text{ Ratio error} = \frac{K_n - R}{R} \times 100 \tag{6.5}$$

where K_n is the nominal ratio for CT, and $R = \frac{\text{actual primary current}}{\text{corresponding secondary current}}$ is the actual ratio.

This ratio error can be reduced by:

1. Using a smaller magnetic path and high-quality core material to lower magnetising and loss components of exciting current.
2. Reducing winding resistance and leakage reactance.
3. Providing turn compensation.

A CT core can support only a finite maximum flux density. When the primary current is so large that the core saturates, flux no longer changes with current; consequently, no secondary current is produced. All primary current then becomes magnetising current. Saturation depends on:

1. Cross-section of the CT core (larger area reduces saturation risk),
2. Quality and type of core steel,
3. Burden connected to the secondary.

The knee-point voltage is the point at which a 10% increase in secondary voltage produces a 50% increase in magnetising current. Beyond the knee point, severe saturation occurs. In differential protection, unequal saturation of two CTs on the same fault current can cause nuisance tripping, as shown in Fig. 6.10.

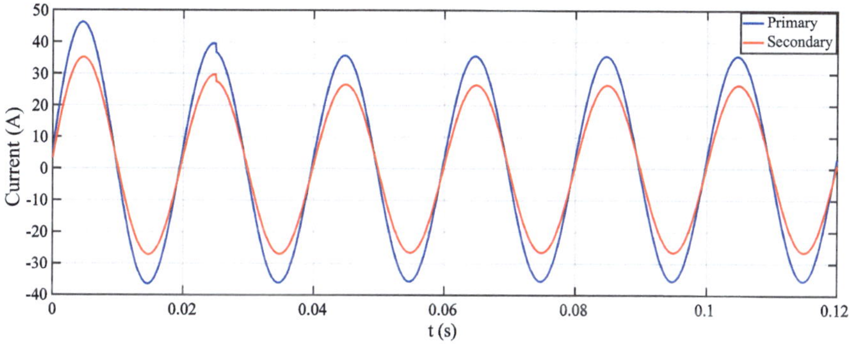

Fig. 6.10 Current waveform with and without saturation

Table 6.1 Standard accuracy classes for CTs

Accuracy class	± % Ratio error at rated current	± Phase displacement error (minutes) at rated current	% Composite error at rated accuracy limit primary current
5P	1	60	5
10P	3	–	10
15P	5	–	15

CTs are classified by accuracy limits specified on the nameplate (e.g., 15 VA, 5P10). The marking indicates the burden and the maximum composite error at 10 times rated current. Standard accuracy classes are shown in Table 6.1.

6.3.2 Potential Transformers

A potential transformer (PT) is a step-down transformer. The primary winding has a higher number of turns, and it is connected across the high-voltage line in parallel. The secondary side of PT has fewer turns and supplies instruments or relays at a safe level, typically rated at 110 or 220 V.

The voltage relationship is:

$$\frac{V_1}{V_2} = \frac{N_1}{N_2}. \tag{6.6}$$

Example 2 A 11,000:110 PT is used with a voltmeter reading 97.5 V. The actual line voltage is:

$$V_1 = 97.5 \times \frac{11\,000}{110} = 9750 \text{ V}.$$

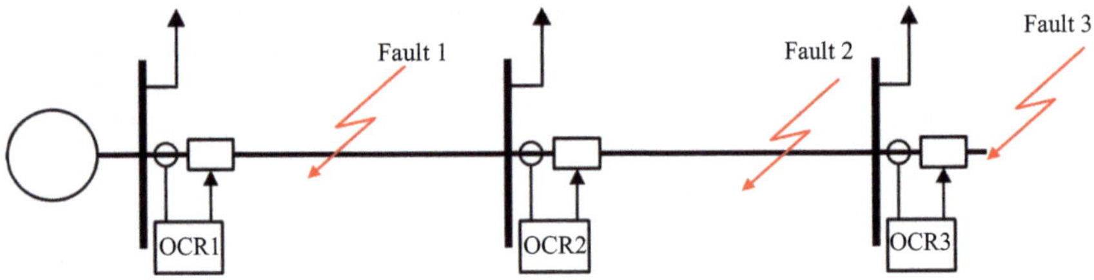

Fig. 6.11 A three-bus example

With the same principle as CTs, PTs cannot be operated as a short circuit.

6.4 Overcurrent Relays

Overcurrent relays (ORs) are among the simplest and most widely applied protective devices in power systems. They continuously monitor the magnitude of current flowing through the transmission lines, detect the fault, and send the trip signal to circuit breakers when the measured current exceeds a pre-set threshold. Their primary function is to detect and isolate faults rapidly while maintaining selectivity to ensure that the faulted section is only disconnected, and leaving the healthy parts of the network in service.

6.4.1 Characteristics of Overcurrent Relays

An overcurrent relay operates when the current magnitude exceeds a fixed pick-up value. It compares the measured current that is obtained from CTs against a threshold and, if the threshold is exceeded, it sends a trip signal to the circuit breaker. For example, as shown in Fig. 6.11, in a radial feeder with three loads, a fault at location Fault 3 causes only the relay OCR3 to operate and trip the breaker supplying the same bus's load. The remainder of the system continues to supply other loads.

6.4.2 Time–Current Curve

A single instantaneous relay cannot provide selectivity in networks with multiple protection stages. Consider a fault at Fault 1 or Fault 2: both produce the same current through relay OCR-2, yet only the downstream relay should operate. Similarly, OCR-2 cannot distinguish between a downstream fault (Fault 2 must trip) and a remote fault (Fault 3 must not trip). The solution is to introduce intentional time grading, or a coordination time interval (CTI), between overcurrent relays. Two principal discrimination methods exist.

Definite-time relays: Each relay is assigned a fixed operating time independent of fault-current magnitude, and the relay nearest the source is given the longest time delay. For a fault at Fault 1, the downstream relay clears the fault before the upstream relays time out. The grading margin must exceed the sum of the downstream breaker interrupting time plus a safety allowance (typically 0.3–0.5 s).

The principal disadvantage is that the relay closest to the source, where fault current is highest, operates with the longest delay, which results in slow fault clearance at the source end.

Inverse-time relays: The operating time is inversely proportional to the fault-current level. This characteristic is obtained by an inverse definite minimum time (IDMT) relay whose time–current curve follows:

$$t = \frac{TMS \times \beta}{(I/I_P)^{\alpha} - 1} \quad (6.7)$$

where t is the operating time, TMS is the time multiplier setting, I is measured current, I_p is the pick-up current, which is typically 125% of the rated current, and α, β define the degree of inversity. The constant values can be selected based on the four standard characteristics defined in Table 6.2.

Where $I_r = I/I_p$. Typically, the TMS for the most downstream overcurrent relay is selected to 0.05 s, and the other relays will have TMS settings according to it. Increasing the TMS moves the TCC curve of relay upward, as shown in Fig. 6.12.

EI relays are particularly suited for distribution feeders supplying loads with high switching inrush, such as refrigerators, pumps, water heaters, and for grading with fuses or auto-reclosers. VI relays excel where fault current decreases substantially with distance from the source; their steep characteristic allows the same TMS to be used for several relays in series while still meeting grading margins.

This very steep inverse behaviour makes EI relays ideally suited for distribution feeder circuits that are subjected to high peak currents on switching-in. Typical loads include refrigerators, pumps, water heaters, and similar appliances that remain connected even after a prolonged supply interruption. The long operating time at normal peak-load values also makes the EI relay particularly suitable for grading with fuses.

Another important application is in conjunction with auto-reclosers in low-voltage distribution circuits. The majority of faults on such systems are transient

Table 6.2 IEC 60,255 defines four standard characteristics

Relay characteristic	Equation (IEC 60255)
Standard Inverse (SI)	$t = \text{TMS} \times \frac{0.14}{I_r^{0.02}-1}$
Very Inverse (VI)	$t = \text{TMS} \times \frac{13.5}{I_r-1}$
Extremely Inverse (EI)	$t = \text{TMS} \times \frac{80}{I_r^2-1}$
Long-time inverse	$t = \text{TMS} \times \frac{120}{I_r-1}$

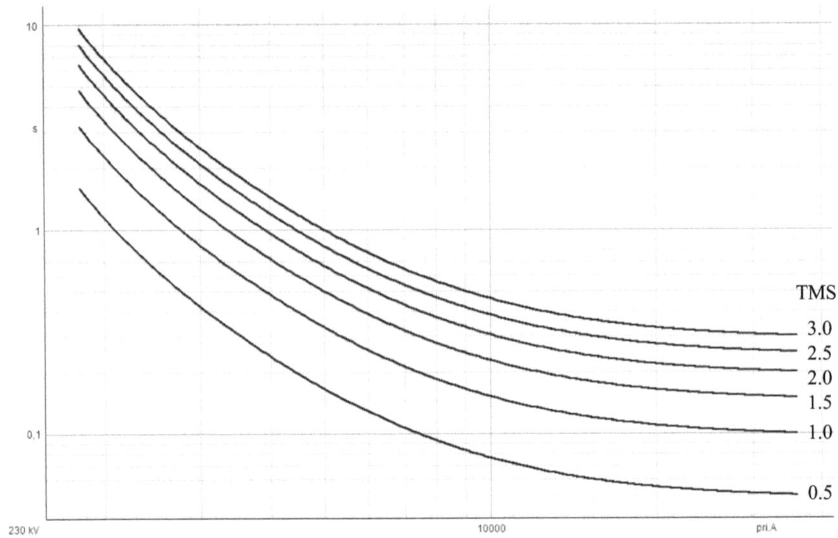

Fig. 6.12 TCC of overcurrent relays for different TMSs

in nature. By setting the auto-recloser to operate before the fuse blows, unnecessary blowing and replacement of fuses in the final circuits can be avoided. If the fault persists after one reclose attempt, the auto-recloser locks in the closed position, and the fuse finally isolates the faulted section. This coordinated operation minimises consumer outages while still providing fast protection for permanent faults.

Compared with the SI curve, the VI curve is much steeper; therefore, the operating time increases much faster for the same reduction in current. This enables the requisite grading margin to be obtained with a lower TMS for the same setting current, thereby minimising the tripping time at the source end of the feeder.

In summary, the choice between EI and VI depends on the system topology and load behaviour:

- Use EI for feeders with high inrush currents, fuse coordination, or auto-recloser schemes.
- Use VI where fault current falls rapidly with distance and uniform TMS grading is required across multiple relays.

Both characteristics complement the basic SI curve and, together with the high-set instantaneous element, allow modern overcurrent protection to achieve fast, selective, and reliable fault clearance across a wide range of distribution and sub-transmission networks. The differences among the operation curves of these characteristics are shown in Fig. 6.13.

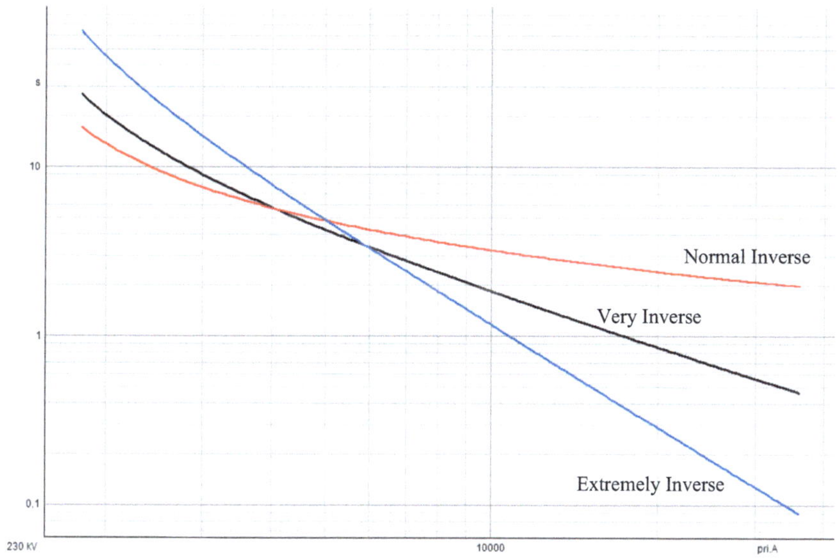

Fig. 6.13 EI, VI, and normal inverse curves of an overcurrent relay

6.4.3 Selecting and Setting the Overcurrent Relays

Two principal settings are required for definite-time or IDMT relays: pick-up current and time delay or TMS.

The pick-up current must be above the maximum load current (including short-time overloads and motor starting) but below the minimum expected fault current. Typical setting:

$$I_p = (1.2 \text{ to } 1.5) \times I_{\text{max load}} \tag{6.8}$$

The Time delay, or grading margin or CTI for upstream relays are:

$$t_{\text{upstream}} = t_{\text{downstream}} + t_{\text{CB}} + \text{security margin} \tag{6.9}$$

The security margin is around 0.3–0.5 s, accounts for relay timing errors, CT errors, and breaker interrupting time.

For TMS, it can be selected initially as 0.05 s, and modified according to the I^2t curve of under protection units.

6.4.4 Coordination Between Two Overcurrent Relays

Correct coordination requires that for any fault location, the backup relay operates after the primary relay with a sufficient grading margin. The operating time of each relay is [1]:

$$t_i = \frac{0.14}{\left(I_{\mathrm{sci}}/I_p\right)^{0.02}} \times \mathrm{TMS}_i(\mathrm{SI\ curve}) \tag{6.10}$$

Coordination criterion:

$$t_{\mathrm{backup}} - t_{\mathrm{primary}} \geq \mathrm{CTI} \tag{6.11}$$

where CTI (coordination time interval) is normally 0.3–0.5 s and equals breaker time + overshoot + safety margin + CT/relay errors.

Example 3 Given CT ratios, rated currents, and maximum short-circuit currents at buses A–D, the relay settings are calculated starting from the farthest relay (TMS = 0.05, pick-up = 1.25 × rated current) and working backwards, adding the CTI at each step. The procedure yields TMS values that satisfy the coordination equation for every fault location.

As a note, insertion of a DG increases fault current contribution. The same relay settings must be re-checked; operating times of both primary and backup relays decrease, but selectivity is usually preserved if the original grading margin was adequate.

6.4.5 Directional Overcurrent Relays (DOCRs)

In ring-main or meshed networks, fault current can flow in either direction through a relay. Non-directional relays cannot discriminate direction. Directional phase-fault overcurrent relays add a directional element that measures the phase angle between voltage and current to determine power-flow direction. Tripping occurs only when current flows in the pre-set "forward" direction and exceeds the overcurrent threshold.

Rules for setting DOCRs:

1. Pick-up current ≥ 1.25 × maximum load current.
2. TMS of the farthest relay from the source = 0.05 s.
3. For meshed systems, directional relays are mandatory.
4. Coordination is verified for a fault located immediately in front of the primary relay (worst-case current for the backup relay).

Examples of DOCR settings in ring-main systems (with given minimum and maximum fault currents at each bus) follow the same stepwise procedure: pick-up calculation, TMS determination from the farthest relay, and verification of CTI for every possible fault location and direction. The resulting settings ensure correct forward operation while blocking reverse faults.

Directional overcurrent protection thus extends the applicability of simple overcurrent relays to complex meshed and ring-main networks while preserving the speed, simplicity, and selectivity essential for reliable power-system operation. The TCC for three overcurrent relays with well coordination is shown in Fig. 6.14.

Example 4 is about how the coordination and settings of relays can be defined.

Example 4 Find the TMS and pick up the current relays R4 and R3, according to the rated current and maximum SC current shown in Table 6.3 based on the Fig. 6.15.

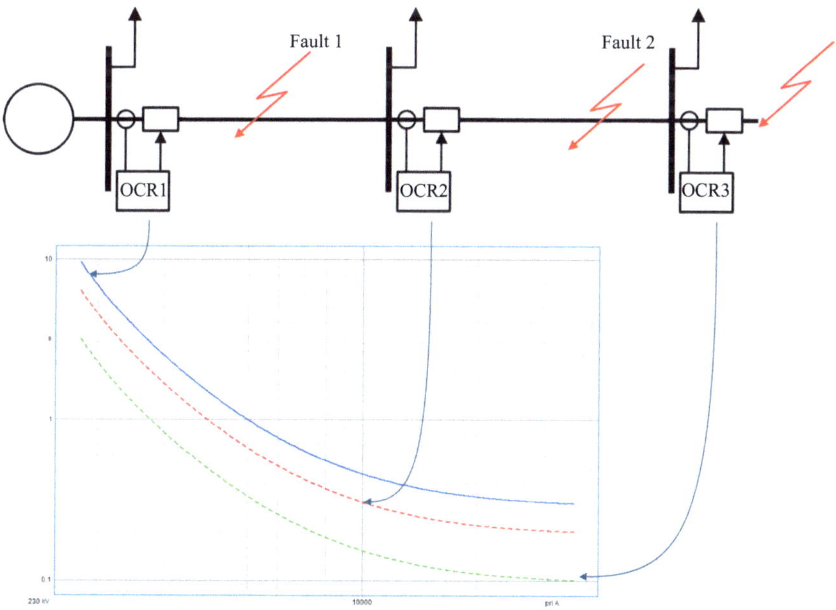

Fig. 6.14 Coordination and TCC curves of overcurrent relays

Table 6.3 Example 4 current values

	Bus A	Bus B	Bus C	Bus D
CT	400/5	400/5	200/5	100/5
Rated current (A)	400	300	150	50
Maximum SC current (A)	7500	5000	2500	1500

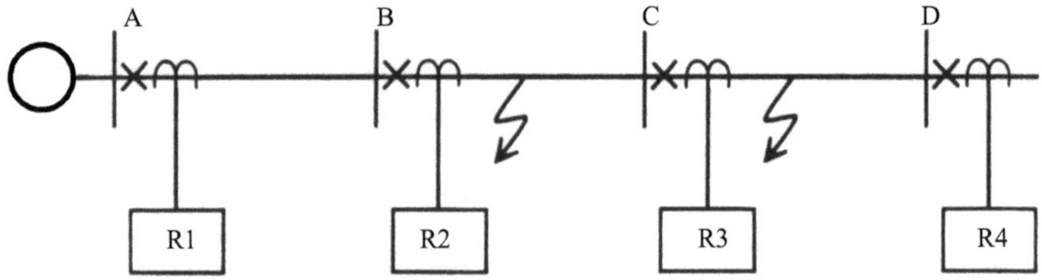

Fig. 6.15 Example 4 diagram

Solution:
First, the pick-up current for relay 4 can be selected:

$$RSI_{R4}(relay.setting.input) = 1.25 \times 50 = 62.5$$

Then, TMS is assumed to be 0.05 s, and the operation time of this relay is determined:

$$t = \frac{0.14}{\left(\frac{I}{I_P}\right)^{0.02} - 1} \times TMS$$

$$t = \frac{0.14}{\left(\frac{1500}{62.5}\right)^{0.02} - 1} \times 0.05 = 0.106$$

Therefore, the next upstream relay should operate after 300 ms of relay 4:

$$t_{R3} = 106 + 300 = 406\,\text{ms}$$

$$t_{R3} = \frac{0.14}{\left(\frac{1500}{150 \times 1.25}\right)^{0.02} - 1} \times TMS = 0.406$$

$$TMS = 0.123$$

6.5 Differential Relays

Differential relays provide the fastest, most selective, and most sensitive protection for any defined portion of the power system. Under normal or external-fault conditions, the algebraic sum of the currents entering and leaving the zone is zero. Any internal fault produces a differential (spill) current that operates the relay. Thus, the task for differential relays is to only operate for internal faults, not externals.

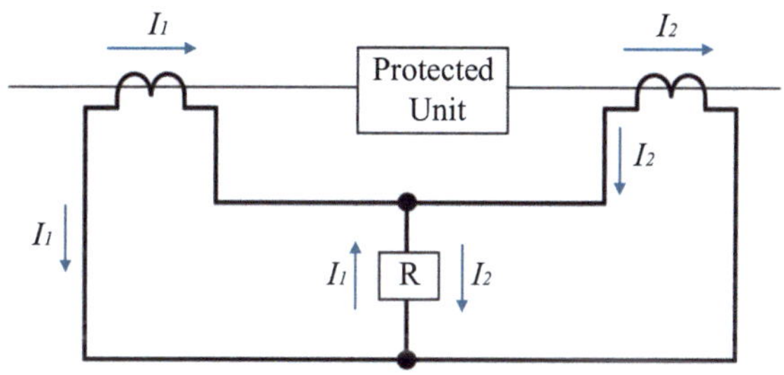

Fig. 6.16 Differential relay installation

6.5.1 Unit Protection Working Principle

In a unit, like a transformer, the fundamental principle is Kirchhoff's current law applied at the zone boundaries:

$$\sum I_{\text{in}} = \sum I_{\text{out}}(\text{healthy or external fault}) \tag{6.12}$$

During an internal fault, the equality is violated, and a differential current appears:

$$I_{\text{diff}} = I_{\text{in}} - I_{\text{out}} \neq 0 \tag{6.13}$$

CTs on each side of the protected zone are connected in opposition so that their secondary currents cancel in the relay operating coil under healthy conditions. The relay therefore receives zero current for external faults or load flow and a net operating current only for internal faults. This gives instantaneous clearance (typically 10–20 ms) confined strictly to the zone defined by CT locations, with no need for time grading or coordination with adjacent protections. A simple diagram of a differential relay is shown in Fig. 6.16. As can be seen, line currents in healthy scenarios should be the same, and the secondary current will also be the same. Therefore, relay seen the 0 A. However, during a fault condition, the relay sees the summation of currents as changing the direction of one of them.

6.5.2 Bias and Saturation

In practice, a non-zero differential current exists even under healthy conditions because of:

- CT ratio mismatch and saturation,
- transformer tap-changer position,
- line charging current,
- magnetising inrush (transformers).

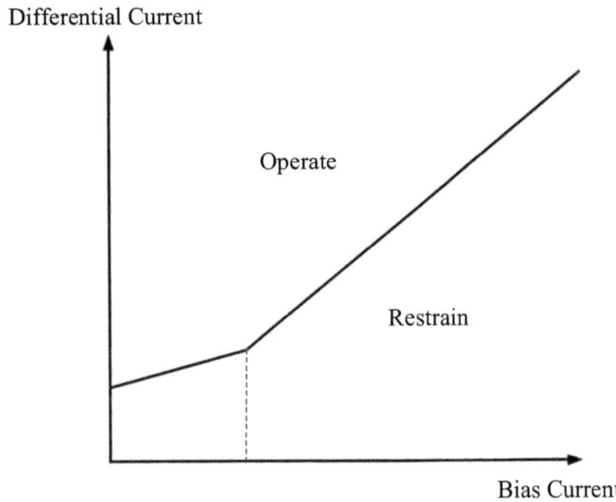

Fig. 6.17 Differential relay operation curve

These currents increase with through-fault magnitude. A plain differential relay set above the maximum possible spill current would be insensitive to low-level internal faults. The classical solution is biased differential protection. A restraint (bias) quantity is derived from the through-current (normally the average or sum of the terminal currents). The relay operates only when

$$I_{\text{diff}} > k \times I_{\text{restraint}} + I_{\text{set}} \tag{6.14}$$

where k is the percentage bias (typically 10–30%), $I_{\text{restraint}} = \frac{|I_1|+|I_2|}{2}$ (or the maximum of the two currents), and I_{set} is the minimum pick-up. This characteristic automatically raises the operating threshold in proportion to the through-current, providing excellent stability for heavy external faults while retaining high sensitivity for internal faults. The bias curve of differential relays is shown in Fig. 6.17.

Differential relay for transformer protection has several advantages, such as:

- High-speed detection of internal faults (one cycle), minimising damage.
- Precise fault location within the zone defined by CT positions.
- No requirement for coordination with other protections.
- Preservation of system stability and reduction of voltage sags.

However, different issues can arise:

- CT ratio mismatch caused by transformer ratio and on-load tap changer.
- Delta–wye transformation of currents (vector-group compensation and zero-sequence elimination).
- Magnetising inrush current (2nd-harmonic blocking or waveform recognition).
- External ground faults on wye windings (zero-sequence current must be blocked or compensated).

When paralleling sources, unequal CT saturation on a through-fault can produce a false differential current; proper CT matching and bias settings are essential.

Differential relays therefore combine absolute selectivity, high speed, and high sensitivity, making them the preferred protection for generators, transformers, busbars, and short lines. The biased and high-impedance variants ensure stability under all external conditions while retaining the ability to detect even low-level internal faults.

6.6 Distance Relays

Distance protection is one of the most widely used schemes for transmission and sub-transmission lines (typically 33 kV and above). The relay continuously measures the apparent impedance $Z = V/I$ at the relay installation location by using the voltage and current signals obtained from the CT and PT. This measured impedance is plotted on the R–X (resistance–reactance) plane and compared with pre-defined operating characteristic curves. If the measured impedance falls inside the protected zone, the relay issues a trip command to the associated circuit breaker.

The principal advantage over overcurrent protection is that the reach of a distance relay is determined almost entirely by the line impedance itself and is independent of the source impedance behind the relay or the contribution from remote infeed. Consequently, the same relay settings can be used for different system operating conditions without losing selectivity. Modern digital/numerical distance relays also exhibit extremely consistent operating times, typically 10–20 ms, regardless of fault current magnitude, unlike older electromechanical or static relays whose speed varied with fault level.

To analyze the distance relay's operation, the power system in Fig. 6.18 can be used. In distance relays, the apparent impedance is calculated by using the current and voltage signals measured at the distance relay location. Analyses in distance relays are highly based on the theory of symmetrical components. The derivation of the impedance equations can be assumed as a radial power system with fault impedance assumed to be zero [2].

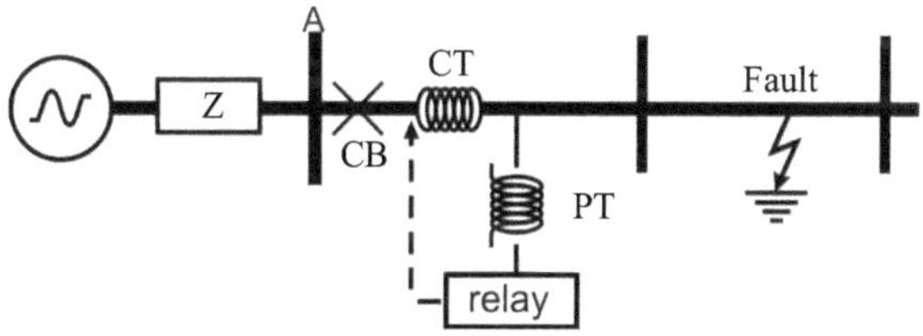

Fig. 6.18 Distance relay installation in lines

During a single phase-to-ground fault, the apparent impedance can be obtained by [3]

$$Z_{AG} = \frac{VR_A}{IR_1 + IR_2 + K_0 IR_0} \tag{6.15}$$

where, Z_{AG} is the apparent impedance, K_0 is the phasor quantity, typically set based on the ratio of Z_0/Z_1 of the line, in which Z_0 and Z_1 are the zero and positive sequence components of line impedance, respectively. Moreover, IR_1, IR_2, and IR_0 are the positive, negative, and zero sequence components of the relay current, respectively.

The main objective of the distance relays is the detection of a zonal fault in a transmission line and sending the trip signal to circuit breakers based on the detected zonal order. Moreover, the zone and time settings should be used for coordination between other distance and overcurrent relays.

6.6.1 Operation Characteristic Curves

Distance relay operation is based on an R–X curve in the complex impedance plane. Several characteristic shapes are used.

6.6.1.1 Plain Impedance Characteristic

The plain impedance characteristic represents the foundational form of distance-relay operating criterion. It is fundamentally a circular boundary centered precisely at the origin of the R–X, resistance–reactance, impedance plane. Any measured apparent impedance vector $\mathbf{Z} = R + jX$ that lies inside or on this circle causes the relay to assert a trip signal. Mathematically, the operating condition is expressed as:

$$|\,\mathbf{Z}\,| \leq Z_r \Rightarrow \sqrt{R^2 + X^2} \leq Z_r \tag{6.16}$$

Or, for computational convenience in digital relays:

$$R^2 + X^2 \leq Z_r^2 \tag{6.17}$$

where Z_r is the relay reach setting, expressed either in primary ohms or secondary ohms after CT/PT ratio compensation. The setting Z_r is chosen to correspond to a desired percentage of the protected line impedance, typically 80–85% for Zone 1, to prevent over-reach due to transient errors.

6.6.1.2 Mho Distance Relays

The Mho relay characteristic is today the most widely deployed distance element in transmission protection worldwide. Unlike the plain impedance circle, the mho characteristic is a circle that passes through the origin of the R–X plane and is

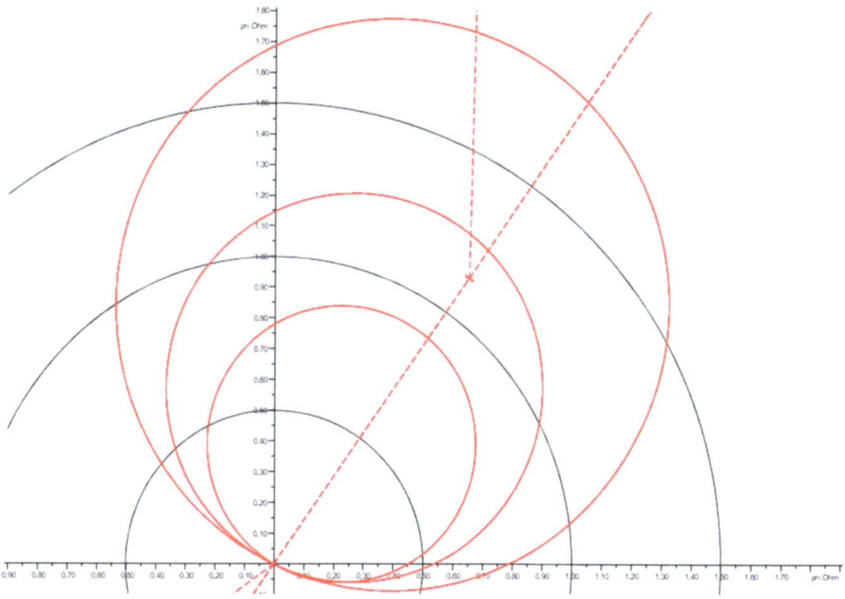

Fig. 6.19 Mho characteristic

tangent to the line impedance locus at the set reach point. This geometry inherently provides excellent directionality while maintaining high resistive-fault coverage. The Mho characteristic is a circle that passes through the origin, giving inherent directionality. The line angle θ and the relay characteristic angle j are defined. The operating equation is

$$| Z |=| Z_s | \cos(\theta - j) \tag{6.18}$$

where Z_s is the relay setting impedance and θ is the measured fault angle. As shown in Fig. 6.19, the Mho R-X plot includes three circles, one for each zone. And each circle has a different operation time, and depending on the fault location, the new impedance seen by the relay will be in one of the circles.

6.6.1.3 Offset Mho/Lenticular Characteristics

There is an issue when the offset Mho relay may operate under maximum load transfer conditions if Zone 3 of the relay has a large reach setting. A large Zone 3 reach may be required to provide remote back-up protection for faults on the adjacent feeder. To avoid this, a shaped type of characteristic may be used, where the resistive coverage is restricted. With a 'lenticular' characteristic, the aspect ratio of the lens (a/b) is adjustable, enabling it to be set to provide the maximum fault resistance coverage consistent with non-operation under maximum load transfer conditions. The lenticular characteristic can tolerate much higher degrees of line loading than offset Mho and plain impedance characteristics.

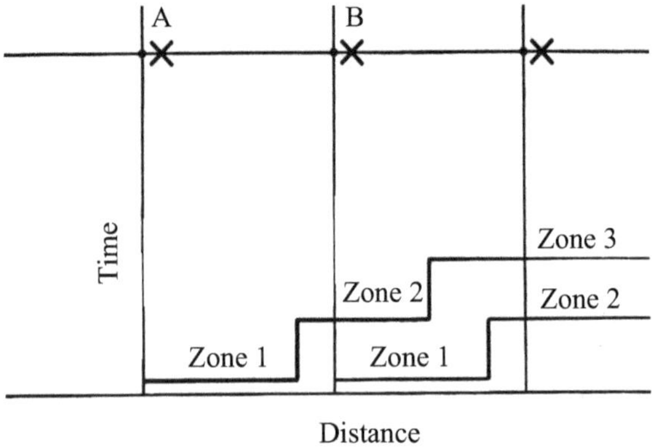

Fig. 6.20 Distance relay zones visualization

6.6.1.4 Reactance Characteristics

A reactance distance relay operates based on the reactive component of fault impedance, with its characteristic represented as a straight line parallel to the Resistance (R) axis on an R-X diagram. It is highly suitable for short transmission lines and resistant to fault arc resistance, but it is non-directional, requiring a separate directional unit to prevent tripping on reverse faults.

6.6.1.5 Quadrilateral Characteristics

The quadrilateral characteristic has become the dominant shape in modern numerical distance relays because of its unmatched flexibility. Unlike plain impedance or mho, the quadrilateral forms a trapezoidal or general polygon on the R–X plane whose resistive and reactive boundaries can be adjusted completely independently. This allows the protection engineer to maximise fault-resistance coverage while excluding load encroachment and power-swing areas. The operating region is defined by four straight-line inequalities:

$$X_{min} \leq X \leq X_{set} \text{ and } R_{left} \leq R \leq R_{right} \tag{6.19}$$

where X_{set} defines the reactive reach, R_{right} sets the maximum resistive reach, R_{left} provides directional supervision, and X_{min} prevents operation for capacitive transients.

6.6.2 Distance Relays Zones

Zone 1 provides the main, high-speed protection for the protected line itself. It is set to cover %80 of the positive-sequence impedance of the line under normal conditions. This conservative reach deliberately leaves the final %20 of the line unprotected by Zone 1 to avoid over-reaching into the next line section due to CT saturation, PT transients, or calculation inaccuracies. Operation time for zone 1 is essentially instantaneous, at 20 ms, with no intentional time delay. For a Mho relay, Zone 1 is implemented as:

$$Z_{r1} = 0.80 \times Z_L \Rightarrow Z_{r1,\text{relay}} = \frac{|Z_{r1}|}{\cos(\phi_L - \phi)} \tag{6.20}$$

where Z_L is the total line impedance, ϕ_L is the line impedance angle, and ϕ is the relay characteristic angle, which is usually set equal to or slightly greater than ϕ_L for security.

Zone 2 extends the protection to cover the end of the protected line and simultaneously acts as the first remote back-up for the adjacent line. Its reach is set to 100% of the protected line impedance plus %50 of the shortest adjacent line or transformer impedance. This overlap guarantees that any fault anywhere on the protected line will be seen inside Zone 2 by the local relay. An intentional time delay of approximately 400 ms is introduced, which is short enough to limit damage and maintain system stability. In equation form for a mho relay:

$$Z_{r2} = Z_L + 0.50 \times Z_{L,\text{next}} \Rightarrow Z_{r2,\text{relay}} = \frac{|Z_{r2}|}{\cos(\phi_{\text{next}} - \phi)} \tag{6.21}$$

Zone 3 serves as the second remote back-up zone, covering the entire next line section and part of the following one. It is typically set to 100% of the protected line plus 100% of the next line plus %25 of the longest third line. The time delay is further increased to 800 ms so that it only operates after both Zone 1 and Zone 2 of the downstream relays have had sufficient time to clear the fault. The setting equation is:

$$Z_{r3} = Z_L + Z_{L,\text{next}} + 0.25 \times Z_{L,\text{next-next}} \Rightarrow Z_{r3,\text{relay}} = \frac{|Z_{r3}|}{\cos(\phi_3 - \phi)} \tag{6.22}$$

In summary, the visualization for each zones operation is shown in Fig. 6.20.

Example 5 Define Zones of distance relay 1. Line 1 is 50 km, Line 2 and 3 are 30 km, and resistances are 0.25 Ω/km, reactances are 0.5 Ω/km. Mho distance relay with a setting angle of 50°, and PT = 66 kV/120 V, CT = 400/5, as shown in Fig. 6.21.

Solution:

Consider a three-line radial system with Line 1 = 50 km and Lines 2 and 3 = 30 km each. Conductor parameters: 0.25 Ω/km, reactances are 0.5 Ω/km. PT ratio = 66 kV/120 V, CT ratio = 400/5, relay characteristic angle = 45°.

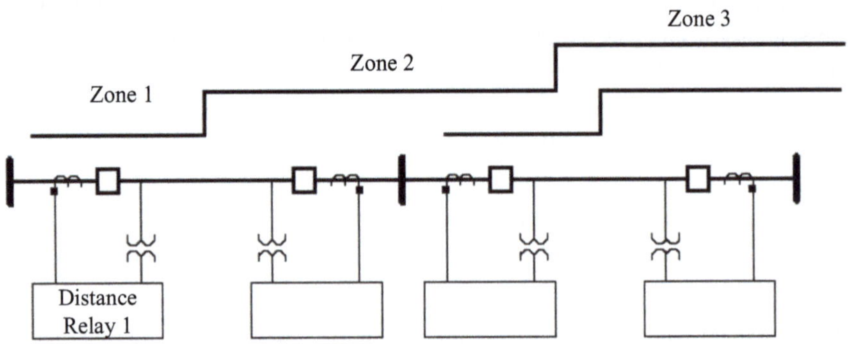

Fig. 6.21 Distance relay example

Zone 1

$$Z_{L1} = 0.8Z_1 = 0.8 \times 50 \times (0.25 + j0.5) = 22.36\angle 63\,\Omega$$

$$Z_{r1} = \frac{22.36}{\cos(63.4^\circ - 50^\circ)} \times \frac{120}{66000} \times \frac{400}{5} = 3.34\angle 50\,\Omega$$

Zone 2

$$Z_{L2} = Z_1 + 0.5Z_2 = 36.34\angle 63.4\,\Omega$$

$$Z_{r2} = \frac{36.34}{\cos(63^\circ - 50^\circ)} \times \frac{220}{66000} \times \frac{400}{5} = 9.95\angle 50\,\Omega$$

Zone 3

$$Z_{L3} = Z_1 + Z_2 + 0.25Z_3 = 48.9\angle 63\,\Omega$$

$$Z_{r3} = \frac{48.9}{\cos(63^\circ - 50^\circ)} \times \frac{220}{66000} \times \frac{400}{5} = 13.42\angle 50\,\Omega$$

6.7 Protection Systems Designing in PowerFactory

6.7.1 Fuse Installation

In PowerFactory, fuses can be installed easily in all cubicles. To start, open an example in PowerFactory and right-click a cubicle to install a fuse, as shown in Fig. 6.22. Click on Devices, Fuse, and New.

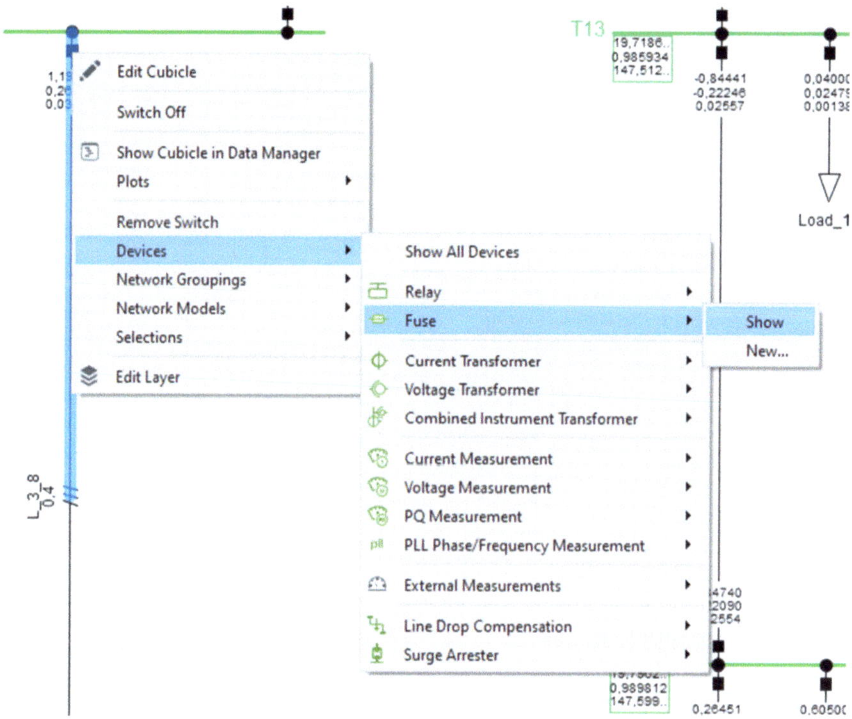

Fig. 6.22 Fuse installation in PowerFactory

Then, the fuse installation window, as shown in Fig. 6.23, will be opened. By selecting the type and type, a fuse from the PowerFactory library can be selected. Then, based on the voltage level of the system, a fuse can be selected. If the voltage level of the power system is not in the library, a higher voltage should be selected. Then, by selecting a fuse, as shown in Fig. 6.24, the fuse is installed.

Then, by right-clicking the same cubicle and selecting Plot > Insert Time-Overcurrent Plot, the fuse's TCC can be observed, as shown in Figs. 6.25 and 6.26. To test the operation of the fuse, after load flow execution or short circuit executing, the plot will be updated, and the operation time will be shown. In the EMT/RMS simulation, the fuse will also open after the specified operation time.

6.7.2 Overcurrent Relay Installation

Relay installation in PowerFactory is also the same as fuse installation. First, right-click the cubicle to install the relay, then select Relay > New, as shown in Fig. 6.27. Then, in the new window, click on the select type, as shown in Fig. 6.28.

Then, in the library of PowerFactory, all relays exist, but a relay with the function of an overcurrent relay should be selected. For doing so, SEL 751 has this

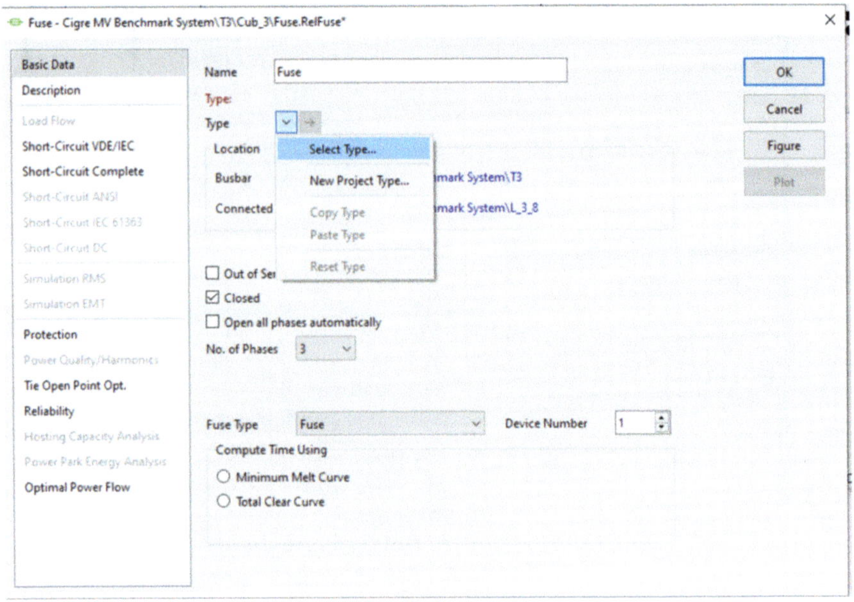

Fig. 6.23 Fuse installation window

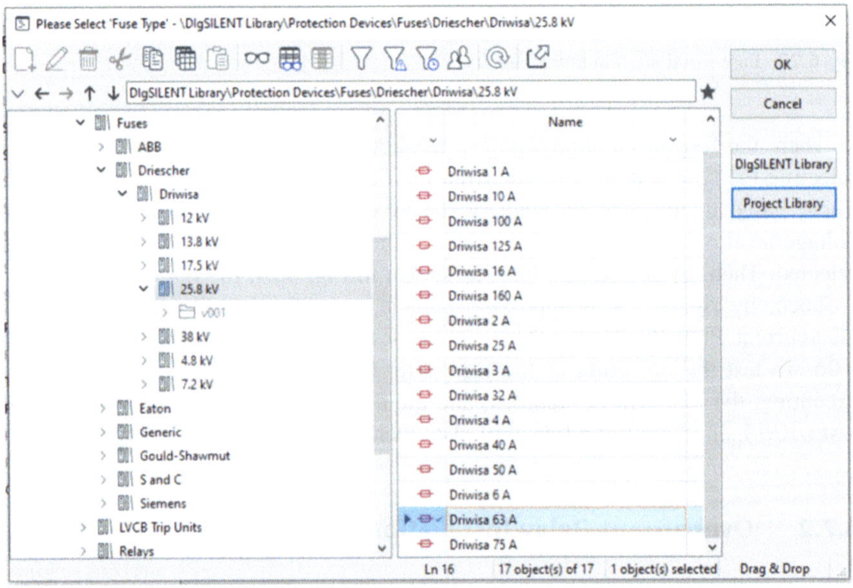

Fig. 6.24 Fuse types in the library

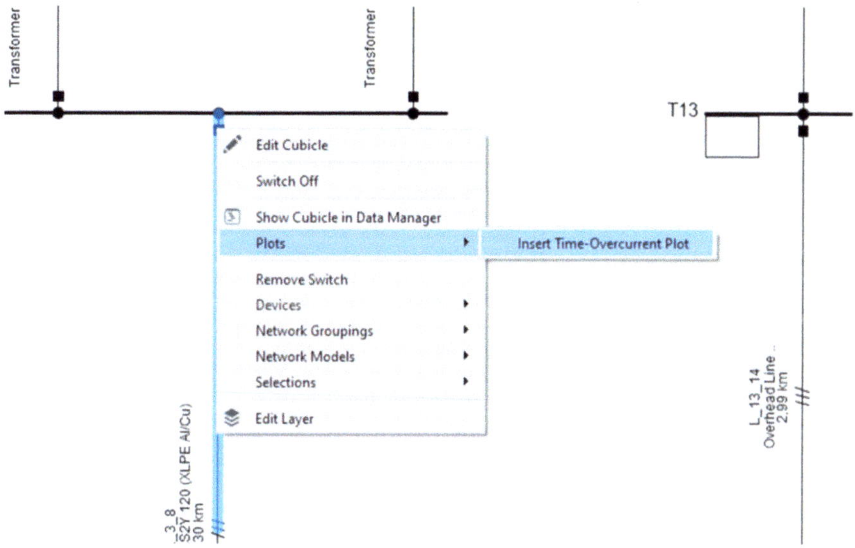

Fig. 6.25 Plotting time overcurrent plot for fuse

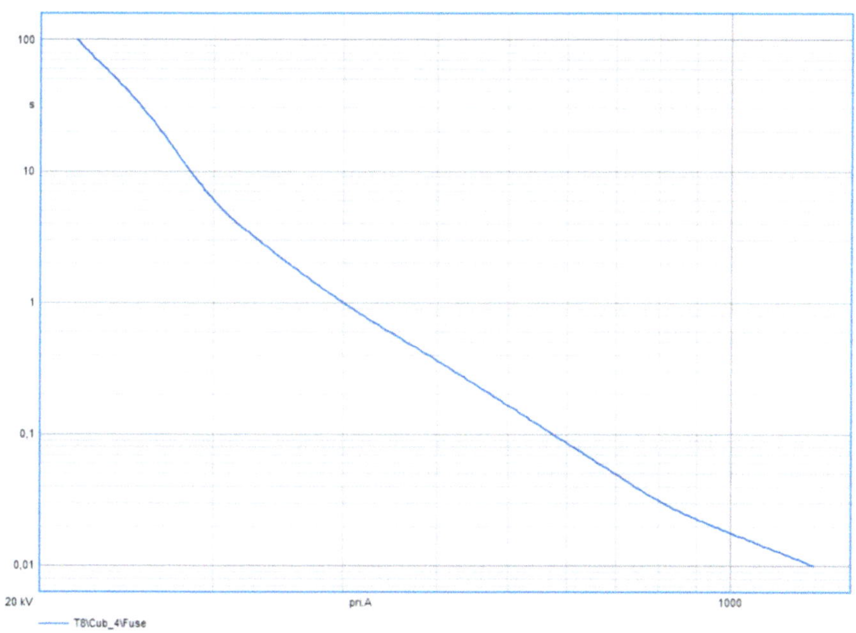

Fig. 6.26 Time-overcurrent plot of the fuse

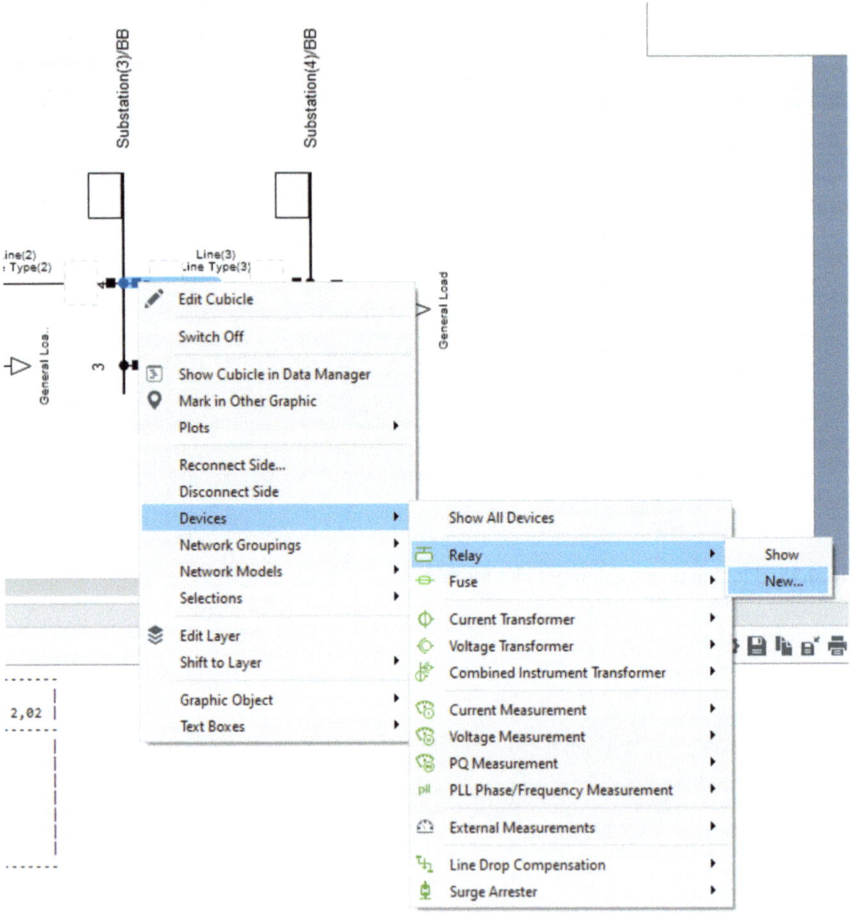

Fig. 6.27 Relay installation

functionality, and can be selected as shown in Fig. 6.29. Also, it can be copied and pasted in the PowerFactory project library to allow for more modifications. Then, the SEL 751 functions are added to the relay, as depicted in Fig. 6.30.

Then, a CT should be installed. By clicking on Create CT, the CT window of Fig. 6.31 is opened, and by making a new type, with 1000:1 A in the next window, as shown in Fig. 6.32, the CT installation will be completed.

By clicking on the contents in the main window of Relay, all unnecessary functions can be disabled. Only 51A, 51B, and 52C, and logic is enough for overcurrent operation. By opening the 51A function, Fig. 6.33, the settings and characteristics of the relay can be inserted. Time dial is the TMS, and the current setting is the pickup current.

Then, with the same method of fuse time overcurrent inserting, the TCC of the overcurrent relay can be observed and tested, as shown in Fig. 6.34.

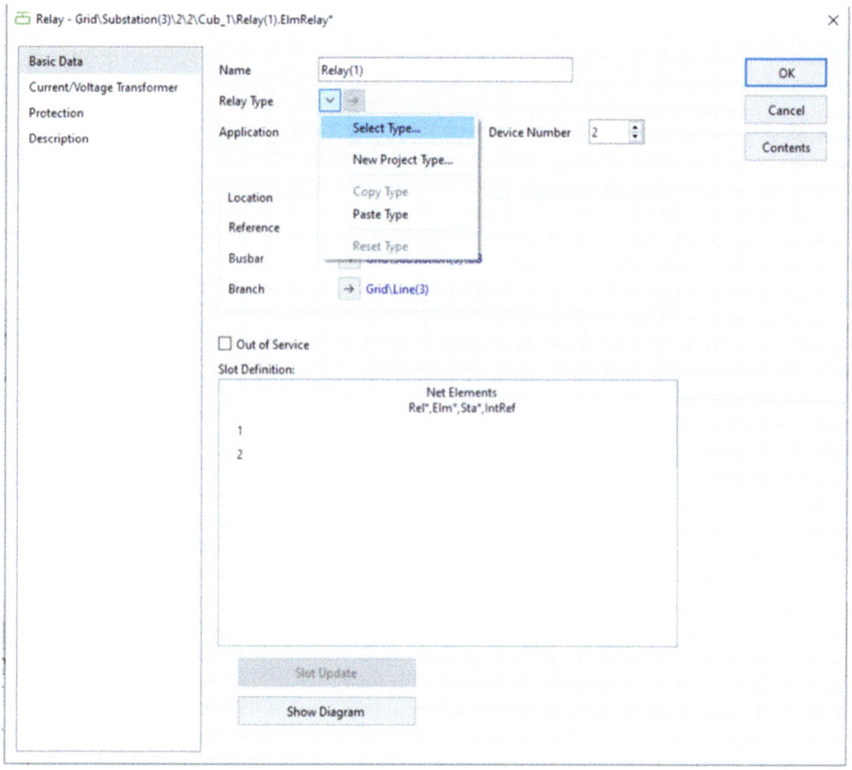

Fig. 6.28 Relay main window

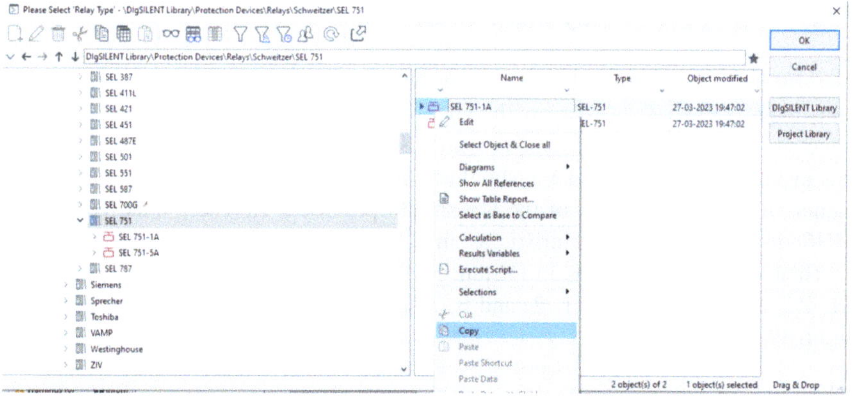

Fig. 6.29 SEL 751-1A selection

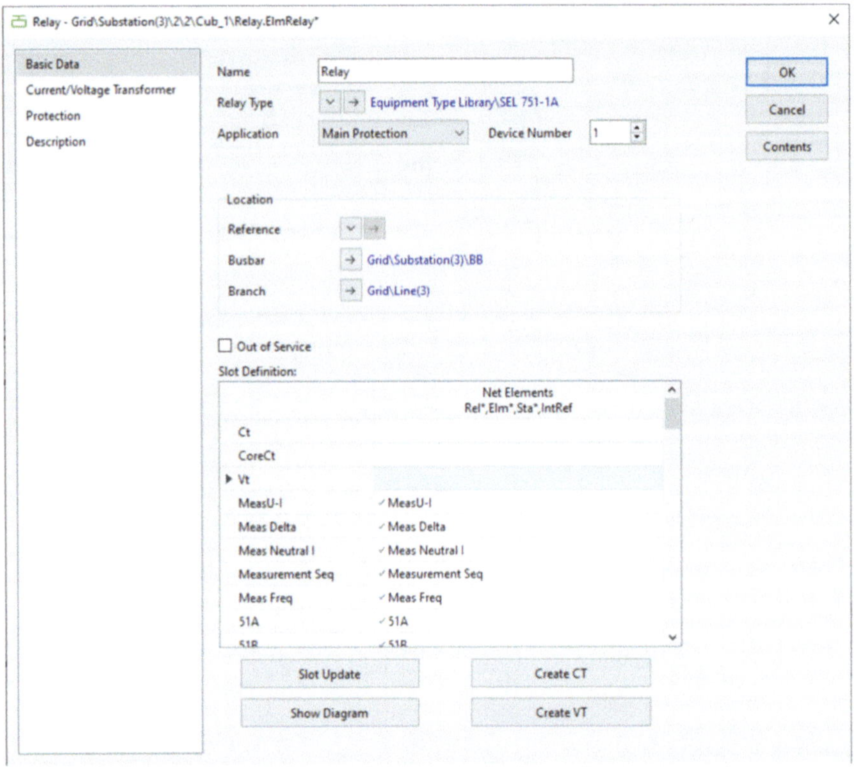

Fig. 6.30 SEL 751 functions

6.7.3 Distance Relay Installation

The distance relay in PowerFactory can be installed in the same way as overcurrent relays. However, a relay with functionality of distance protection should be selected. In this book, the Siemens relay 7SA522 5A, as shown in Fig. 6.35. After this installation, the CT and VT should be installed according to the relay requirements; thus, the secondary of the CT should provide 5 A. Then, making only the MHO distance relay characteristic as an active function, as shown in Fig. 6.36,

Then, from the functions in Distance Mho, the zone values can be provided in Z1, Z2, and Z3 for zones 1, 2, and 3, respectively, as shown in Figs. 6.37, 6.38 and 6.39.

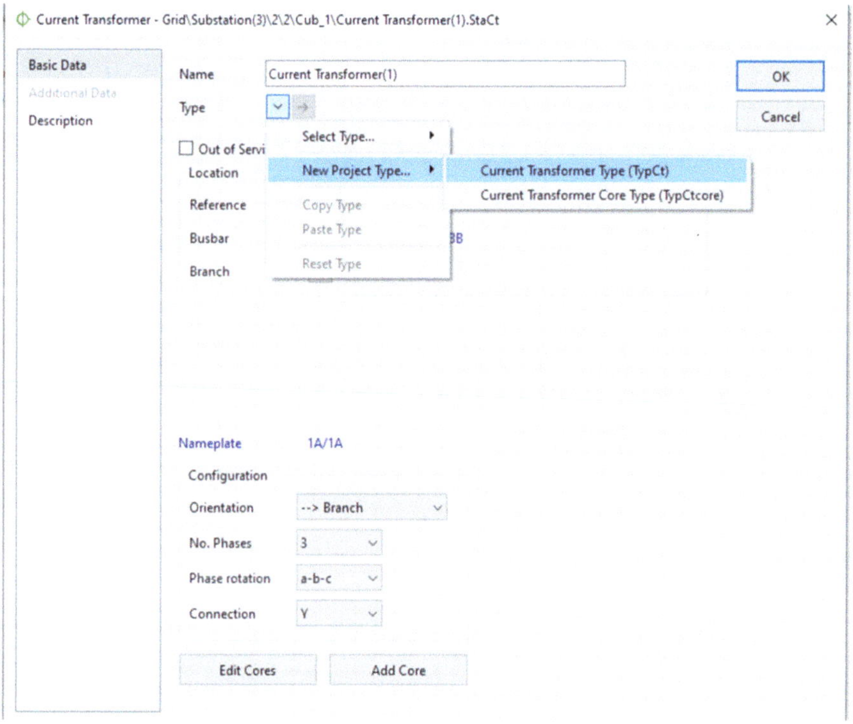

Fig. 6.31 Create CT

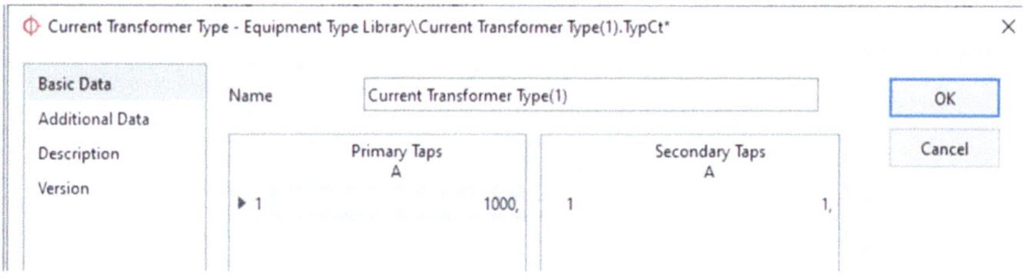

Fig. 6.32 CT ratings

6.8 Summary

This chapter provides a complete and practical introduction to the main protection devices used in power systems. It explains how these devices work together to detect faults quickly, isolate only the faulty part of the network, and protect expensive equipment such as generators, transformers, and transmission lines from serious damage. The chapter begins with fuses, describing their simple construction, fast operation under high currents, time–current behaviour, and the important

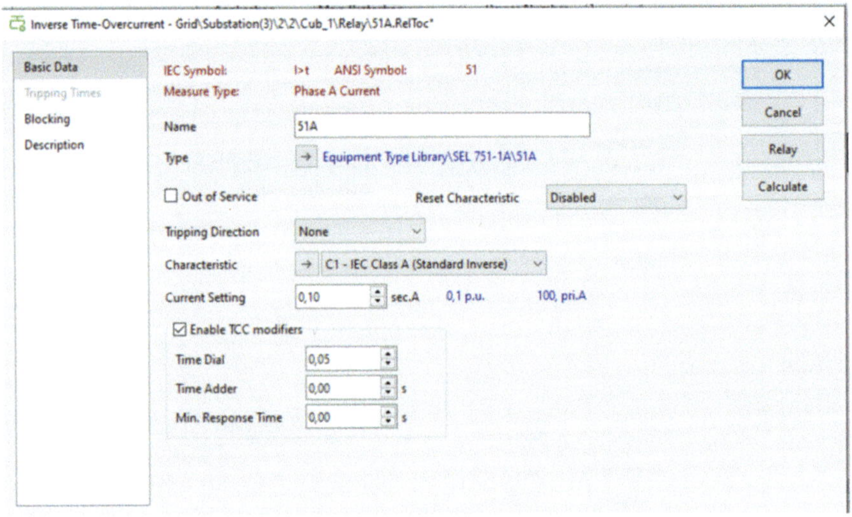

Fig. 6.33 51A settings for overcurrent relays

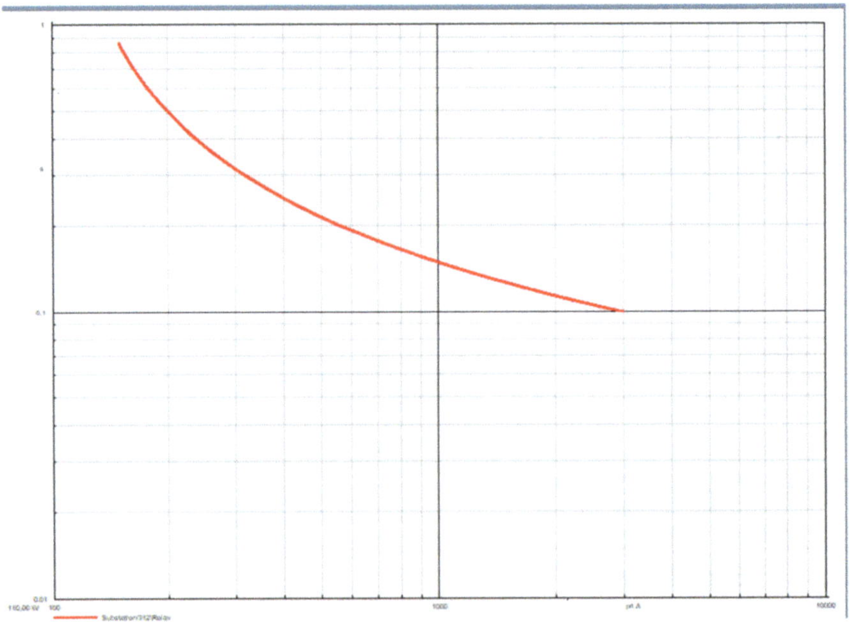

Fig. 6.34 TCC of the overcurrent relay in PowerFactory

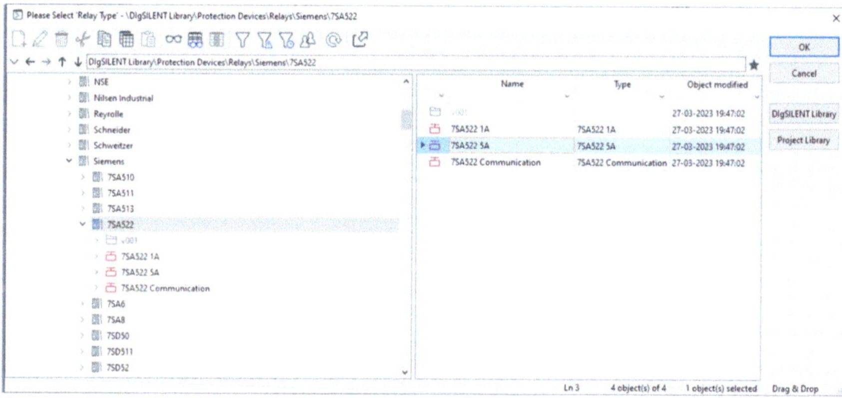

Fig. 6.35 Distance relay installation in PowerFactory

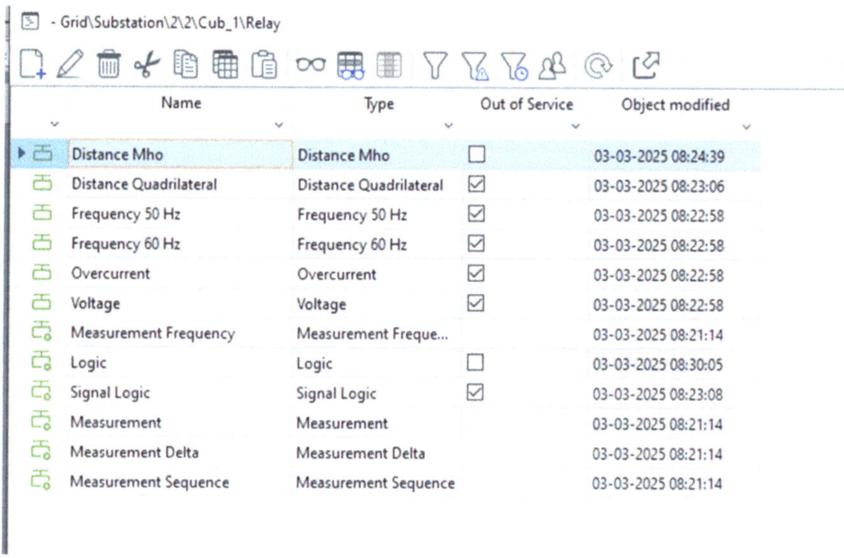

Fig. 6.36 Distance relay functions

rules for coordinating them in distribution networks so that only the fuse closest to the fault operates. Overcurrent relays are discussed in detail, including their different operating characteristics, how settings are selected, and the step-by-step coordination process between primary and backup relays. Differential protection is presented as one of the fastest and most selective methods, with emphasis on why bias is used and how it overcomes practical challenges like transformer inrush and measurement errors. Distance protection explained by the main operating shapes, the concept of three protection zones with different reaches and time

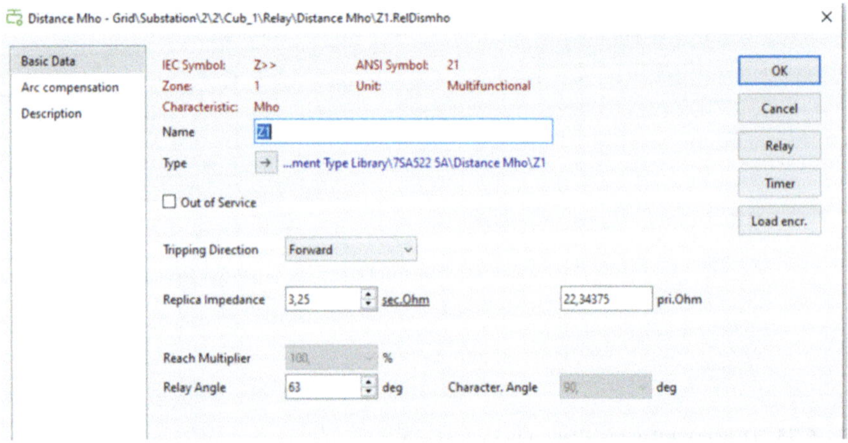

Fig. 6.37 Zone 1 setting

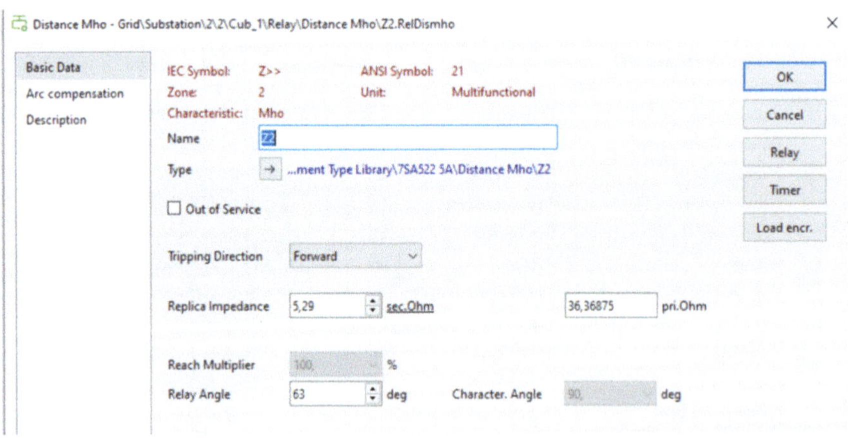

Fig. 6.38 Zone 2 setting

delays, and the way engineers balance speed and backup capability. The chapter ends with clear, step-by-step guidance on implementing all these devices in PowerFactory, including fuse installation, overcurrent relay settings, and full distance relay configuration using real examples.

6.9 Problems and Questions

1. For a 250:5 CT with ammeter reading 4 A, derive the actual primary current using the transformation ratio equation $I_1/I_2 = N_2/N_1$.

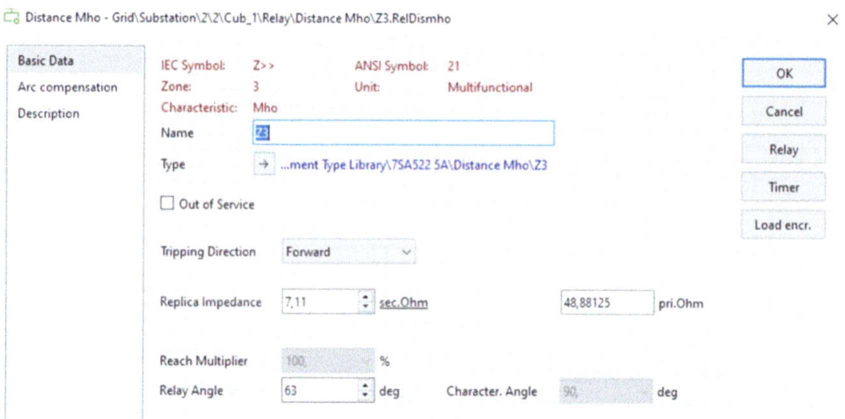

Fig. 6.39 Zone 3 setting

2. An overcurrent relay uses the standard inverse characteristic. Given $I_p = 8$ A, TMS $= 0.1$, find the operation time.
3. Two overcurrent relays are coordinated with CTI $= 0.4$ s. If downstream $t_p = 0.25$ s, derive the required upstream operating time.
4. For a biased differential relay, the operating condition is $| I_d | > 0.2 | I_r | + 0.05 I_n$. If $I_1 = 10$ pu and $I_2 = 9.7$ pu, calculate I_d and I_r and determine if the relay operates.
5. A distance relay Zone 1 is set to 80% of line impedance, $Z_L = 15 + j35\ \Omega$. Write the secondary reach equation after CT $= 1000/5$ and VT $= 220$ kV/110 V conversion.

References

1. Bayati N, Dadkhah A, Sadeghi SH, Vahidi B, Milani AE (2017) Considering variations of network topology in optimal relay coordination using time-current-voltage characteristic. In: 2017 IEEE international conference on environment and electrical engineering and 2017 IEEE Industrial and commercial power systems Europe (EEEIC/I&CPS Europe). IEEE, pp 1–5
2. Bayati N (2022) Novel modified characteristic curve of distance relays by consideration of harsh weathers. In: 2022 IEEE international conference on environment and electrical engineering and 2022 IEEE industrial and commercial power systems Europe (EEEIC/I&CPS Europe) 2022 Jun 28. IEEE, pp 1–4
3. Ziegler G (2011) Numerical distance protection: principles and applications. John Wiley & Sons

Stability and Control of Power Systems

7

Abstract

This chapter has presented the fundamental concepts of power system stability and control, which are essential for ensuring reliable, secure, and continuous operation of modern electric grids. It began by classifying stability into voltage stability and frequency stability, highlighting how disturbances such as short circuits, large load changes, or generator outages can lead to unacceptable voltage drops, frequency excursions, or loss of synchronism. The dynamics of synchronous machines were then examined through the swing equation, the power-angle characteristic, and the single-machine infinite-bus model, providing the theoretical basis for understanding rotor-angle behavior under both small and large disturbances. Steady-state stability was discussed using the synchronizing power coefficient, while transient stability was analyzed with the equal-area criterion in both the power-angle and power-time domains. These analytical tools enable engineers to determine critical clearing angles and times without full numerical integration of the swing equation. The chapter then moved to practical control strategies. Active-power and frequency control were structured hierarchically: primary (governor droop) for rapid stabilization, secondary (AGC via area control error) for frequency restoration and tie-line scheduling, and tertiary for economic dispatch and reserve management. The power–frequency characteristic of interconnected systems and the impact of reduced system inertia due to inverter-based resources were emphasized. Reactive-power and voltage control were addressed next, covering the generation and absorption of reactive power by generators, transmission lines, and loads. Voltage regulation methods were presented in detail, including static shunt and series capacitors, synchronous compensators, SVCs, and STATCOMs, with phasor diagrams and governing equations illustrating their operating principles and relative advantages. Finally, a complete step-by-step PowerFactory tutorial demonstrated the implementation of governor and excitation systems on a single-machine example, showing how frequency and voltage remain stable after a large load-switching disturbance when proper controllers are applied.

N. Bayati and M. A. Khan, *Power System Fundamentals with PowerFactory*,
https://doi.org/10.1007/978-3-032-32450-4_7

Keywords

Stability • Inertia • Frequency control • Voltage control

7.1 Introduction

The reliable operation of a power system requires that the system maintain the voltage, frequency, and synchronism among generators at an acceptable level during both normal conditions and following disturbances. Power system stability refers to the ability of the interconnected power system, for a given initial operating condition, to regain a state of operating equilibrium after being subjected to a physical disturbance, such as a short circuit or large load switching, with all system variables bounded so that practically the entire system remains intact.

Disturbances in power systems can range from small load changes and switching operations to severe events such as short circuits, loss of a large generator, or sudden loss of a major transmission line. In each case, the system must exhibit sufficient restoring forces to prevent progressive deterioration that could lead to loss of synchronism, unacceptable frequency excursions, or widespread voltage collapse.

Stability in power systems can be divided into different types, including transient stability, steady stability, and dynamic stability, as defined in Fig. 7.1.

Power system stability is traditionally classified into two fundamental categories:

- Voltage stability: The ability of the grid to maintain steady voltages at all buses following a disturbance. Voltage instability typically manifests as a progressive fall or rise of voltages and may lead to loss of load or cascading line tripping. Depending on the grid codes, typically voltages can drop only around 3–5%.

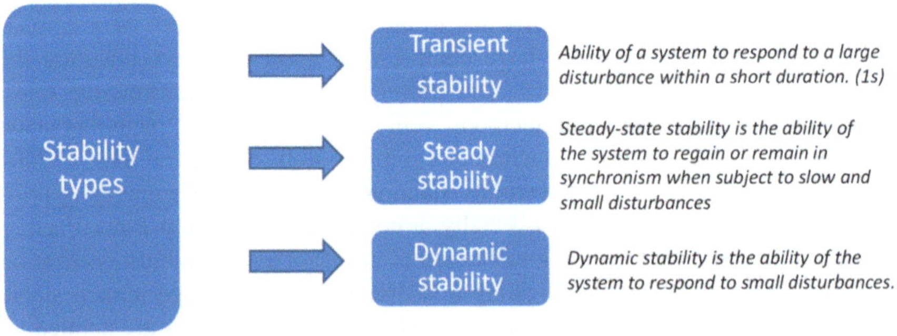

Fig. 7.1 Stability types in power systems

- Frequency stability: The ability to maintain steady frequency following severe disturbances that result in a significant imbalance between generation and load. Frequency instability can lead to sustained frequency deviations that trigger under- or over-frequency protection, load shedding, or generator tripping. Depending on the grid codes, mandatory power frequency response is triggered below 49.5 Hz, with time-limited operations allowable down to 47.5 Hz or 47.0 Hz, depending on the region.

After a short circuit, as shown in Fig. 7.2, without proper control, the voltage will drop to around zero, and the frequency will increase or decrease to an unstable value. This situation is not acceptable for the grid to be operated.

In modern power systems, stability and control have become significantly more challenging due to the rapid integration of renewable energy sources and the corresponding displacement of conventional synchronous generation. Inverter-based resources (IBRs) contribute limited or no rotating inertia, reduced short-circuit levels, and fundamentally different dynamic behavior compared to synchronous machines. This transition reduces the natural damping and synchronizing torque available in the system, increases the rate of change of frequency (RoCoF) after disturbances, and changes the nature of voltage and frequency support during faults and recovery periods.

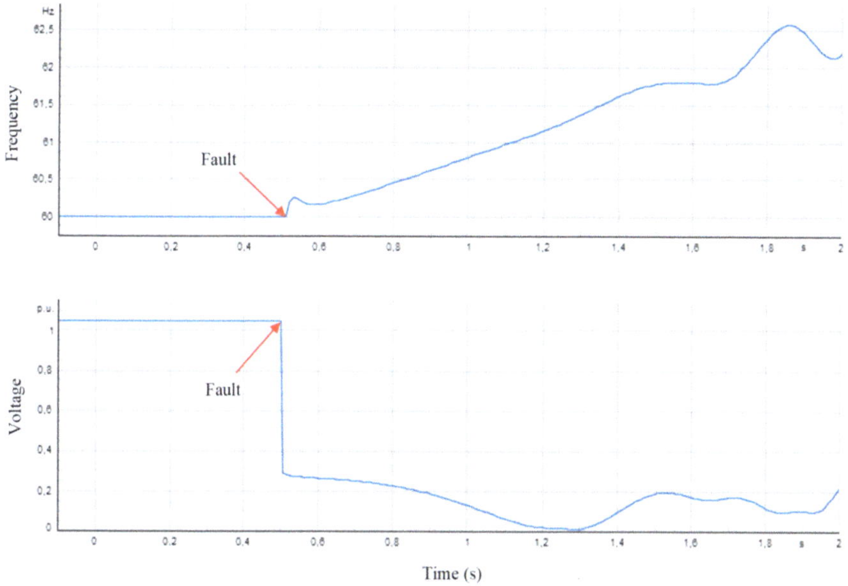

Fig. 7.2 Frequency and voltage instability due to the fault

7.2 Stability in Power Systems

7.2.1 Synchronous Machine Dynamics

A large-scale power system consists of numerous synchronous generators operating in parallel and interconnected through transmission lines. Under normal conditions, all generators are rotating at the same electrical speed, synchronous speed, or frequency to maintain a constant relative angular position, which we call it: they are in synchronism.

When the system experiences a disturbance, for example, a short circuit, line tripping, sudden load change, or loss of generation, electromagnetic and mechanical torques acting on each rotor become temporarily unbalanced. This imbalance causes acceleration or deceleration of individual machines, and then, according to Newton's second law for rotation [1]:

$$J\frac{d^2\delta}{dt^2} = T_m - T_e \tag{7.1}$$

where, J = moment of inertia of the rotor (kg m^2), δ = rotor angle (electrical radians), T_m = mechanical torque from the prime mover (N m), T_e = electromagnetic torque developed by the machine (N m), The key dynamic variable is the rotor angle δ, which represents the angular displacement of the rotor magnetic axis relative to a synchronously rotating reference frame. If δ increases continuously without bound or oscillates with increasing amplitude, the machine loses synchronism, leading to transient instability.

7.2.2 Pattern of Stability Curves

The steady-state power transferred between a synchronous generator and the rest of the grid depends on the rotor angle δ. Typically, the rest of the grid is shown by an infinite grid or busbar, as shown in Fig. 7.3, which means the oscillations in the generator cannot impact the rest of the grid. Then, the electrical power output follows the well-known power-angle curve, called the power-angle characteristic:

$$P_e = \frac{EV}{X}\sin\delta \tag{7.2}$$

Where, E= internal voltage magnitude behind synchronous reactance (per unit), V= terminal (or infinite bus) voltage magnitude (per unit), X= total transfer reactance between internal voltage and system bus (per unit). From this equation, we can understand that the maximum power transfer occurs at $\delta = 90°$, for stable operation, the operating point must lie on the rising part of the curve ($\delta < 90°$).

Small disturbances cause the rotor to oscillate around the equilibrium point. The frequency and damping of these oscillations depend on the synchronizing torque coefficient ($\partial P_e/\partial\delta$) and damping torque. However, large disturbances can drive δ rapidly toward or beyond the unstable region, requiring transient stability analysis.

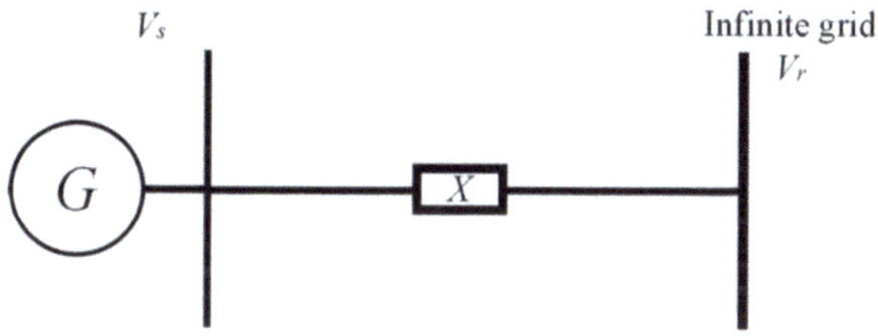

Fig. 7.3 Single machine power system

7.2.3 Single Machine with Infinite Bus

The single-machine infinite-bus system is the classical model used to study the fundamental electromechanical behavior and transient stability of synchronous generators in large interconnected power systems. In this simplified representation, a single synchronous generator is connected to a very large power system through a transmission line, as shown in Fig. 7.3. The infinite bus is idealized as having constant voltage magnitude and constant frequency, unaffected by the power exchange with the generator.

The generator has an internal voltage source of $E\angle\delta$ with synchronous reactance of X_d, and it is connected by a lossless transmission line with total equivalent reactance of X_{TL} to the infinite bus with a voltage of $V\angle 0°$. The angle δ is the rotor angle, defined as the angular displacement between the generator's internal voltage phasor and the infinite bus voltage phasor.

$$\begin{aligned} V_s &= E\angle\delta \\ V_r &= V\angle 0° \end{aligned} \tag{7.3}$$

The equivalent circuit of this system is shown in Fig. 7.4. Since the transmission line is lossless, the active power delivered by the generator equals the active power received at the infinite bus. The line current is given by:

$$I_s = \frac{E\angle\delta - V\angle 0°}{jX} \tag{7.4}$$

where, $X = X_d + X_{TL}$ is the total transfer reactance between the internal voltage and the infinite bus.

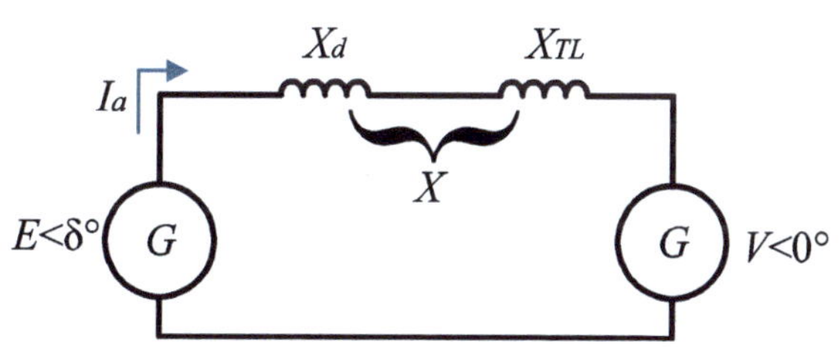

Fig. 7.4 Equivalent circuit of a grid-connected single machine

The complex power delivered by the generator is:

$$\begin{aligned} S &= P_s + jQ_s = V_s I_s^* \\ P_s + jQ_s &= j\frac{E^2}{X} + \frac{EV}{X}\sin\delta - j\frac{EV}{X}\cos\delta \end{aligned} \tag{7.5}$$

After substitution and simplification, the active and reactive power expressions become:

$$\begin{aligned} P_s &= \frac{EV}{X}\sin\delta \\ Q_s &= \frac{E^2}{X} - \frac{EV}{X}\cos\delta \end{aligned} \tag{7.6}$$

For lossless transmission lines, the sending power is equal to the receiving end power. In case of three-phase system, this power can be written as,

$$P_e = P_s = P_r = 3\frac{EV}{X}\sin\delta \tag{7.7}$$

where the maximum transferable power is:

$$P_{\max} = 3\frac{EV}{X} \tag{7.8}$$

Or:

$$P_e = P_{\max}\sin\delta \tag{7.9}$$

Therefore, for any conditions, we have:

$$P_0 = P_{\max}\sin\delta_0 \tag{7.10}$$

$$\sin\delta_0 = \frac{P_0}{P_{\max}} \tag{7.11}$$

$$\delta_0 = \sin^{-1}\left(\frac{P_0}{P_{\max}}\right) \tag{7.12}$$

Example 1 The terminal voltage of a three-phase wye-connected synchronous generator is 33 kV. The generator is connected to an infinite bus through a transmission line. The reactance of the generator and transmission line is 1 Ω and 0.6 Ω, respectively. The generator delivers the power of 50 MW to the infinite bus, and the infinite bus voltage is found to be 11 kV. Calculate the torque angle, transmission line current, and three-phase reactive power consumed by the transmission line.

Solution: Per phase synchronous generator voltage is,

$$E_f = \frac{33}{\sqrt{3}} = 19052.55\,\text{V}$$

The per-phase voltage at the infinite bus is calculated as,

$$V_{\text{ib}} = \frac{11}{\sqrt{3}} = 6350.85\,\text{V}$$

The value of the torque angle can be determined as,
Given: $P = 50$ MW, $X = 0.6 + 1 = 1.6\ \Omega$ (total reactance), voltages in kV line-to-line

$$50 = 3 \cdot \frac{33 \times 11}{\sqrt{3} \times \sqrt{3}} \cdot \frac{1}{1.6} \sin\delta$$

$$\delta = 12.7^\circ$$

The value of the current in the transmission line can be calculated as,

$$I_L = \frac{19052.55\angle 12.7^\circ - 6350.85\angle 0^\circ}{1.6\angle 90^\circ} = 8082.92\angle -71.1^\circ\,\text{A}$$

The value of the three-phase reactive power consumed can be calculated as,

$$Q = 3I_L^2 X_L = 3 \times (8082.92)^2 \times 1.6 = 313\,\text{Mvar}$$

7.2.4 Steady-State Stability

Steady-state stability is the ability of the power system to remain in synchronism when subjected to small and gradual disturbances. Some examples for these typical disturbances are slow changes in load demand, minor variations in generation output, small adjustments in voltage set points, and gradual switching of shunt reactors or capacitors.

These changes are slow enough that the system dynamics remain close to equilibrium at all times, and the rotor angles deviate only slightly from their pre-disturbance values.

In contrast to transient large-disturbance stability, steady-state stability is concerned with the inherent restoring forces present in the system under small perturbations. The key mechanism providing this restoring action is the synchronizing torque that acts to return the rotor to its equilibrium position whenever a small angular deviation occurs.

From the power-angle characteristic derived in the single-machine infinite-bus model,

$$P_e = P_{\max} \sin\delta, P_{\max} = \frac{3EV}{X} \tag{7.13}$$

the change in electrical power for a small change in rotor angle is given by the slope of the curve:

$$\frac{\Delta P_e}{\Delta\delta} \approx \frac{\partial P_e}{\partial\delta} = P_{\max} \cos\delta = \frac{3EV}{X} \cos\delta \tag{7.14}$$

This quantity is called the synchronizing power coefficient. The corresponding synchronizing torque coefficient is:

$$K_s = \frac{\partial T_e}{\partial\delta} = \frac{P_{\max}}{\omega_s} \cos\delta \tag{7.15}$$

where ω_s is the synchronous speed in rad/s.

The sign and magnitude of K_s determine the nature of the response to a small disturbance:

- Positive K_s ($\cos\delta > 0$, i.e., $\delta < 90°$): An increase in rotor angle produces an increase in electrical power, which decelerates the rotor back toward the equilibrium point, leading to a stable condition.
- Negative K_s ($\delta > 90°$): An increase in angle produces a decrease in electrical power and further acceleration, leading to an unstable condition.

Thus, the steady-state operating point is stable only when the system operates on the rising portion of the power-angle curve ($\delta < 90°$).

Example 2 A generator works under P.F = 0.8, X = 0.7 pu, and is connected to an infinite bus. Find Delta, E (generator voltage), Q, and P.

Solution:

$$\bar{V} = 1\angle 0° \text{ pu}$$

$$\cos\varphi = 0.8 \Rightarrow \varphi = \arccos(0.8) = 36.9°$$

$$\bar{I} = 1\angle -36.9° \text{ pu} = 0.8 - j0.6 \text{ pu}$$

$$\bar{E}_0 = jX_s\bar{I} + \bar{V}$$

$$\bar{E}_0 = j0.7(0.8 - j0.6) + 1$$

$$= j0.7 \cdot 0.8 - j^2 0.7 \cdot 0.6 + 1 = 0.56 + 0.42 + 1 + j0.56 = 1.42 + j0.56$$

$$E_0 = \sqrt{1.42^2 + 0.56^2} = \sqrt{2.0164 + 0.3136} = \sqrt{2.33} \approx 1.53\,\text{pu}$$

$$\delta_0 = \tan^{-1}\left(\frac{0.56}{1.42}\right) \approx 21.5^\circ$$

$$\bar{E}_0 = 1.53\angle 21.5^\circ \text{ pu}$$

$$P_0 = \frac{E_0 V}{X_s}\sin\delta_0 = \frac{1.53 \times 1}{0.7}\sin 21.5^\circ \approx 0.8 \text{ pu}$$

$$Q_0 = \frac{E_0 V}{X_s}\cos\delta_0 - \frac{V^2}{X_s} = 0.6 \text{ pu}$$

7.2.5 Equal Area Criterion

The equal area criterion is a powerful graphical method for determining the transient stability of a synchronous machine connected to an infinite bus following a large disturbance, without having to solve the full swing equation numerically. It is based on the principle of conservation of energy and provides a clear physical interpretation of acceleration and deceleration during rotor swings.

During a disturbance, the electrical power output P_e changes suddenly while the mechanical power input P_m from the prime mover remains essentially constant. The difference between these two powers is the accelerating power, which causes the rotor to accelerate or decelerate according to the swing equation:

$$\frac{2H}{\omega_s}\frac{d^2\delta}{dt^2} = P_m - P_e \tag{7.16}$$

Multiplying both sides by $d\delta/dt$ and integrating gives the energy balance form:

$$\int_{\delta_0}^{\delta} (P_m - P_e)d\delta = \frac{H}{\omega_s}\left(\left(\frac{d\delta}{dt}\right)^2 - 0\right) \tag{7.17}$$

The left-hand side represents the net change in kinetic energy of the rotor. For the machine to remain stable after the first swing, the rotor must reach a maximum angle δ_m where its speed deviation returns to zero ($d\delta/dt = 0$). At this point, the total accelerating area must be exactly equal to the total decelerating area that follows.

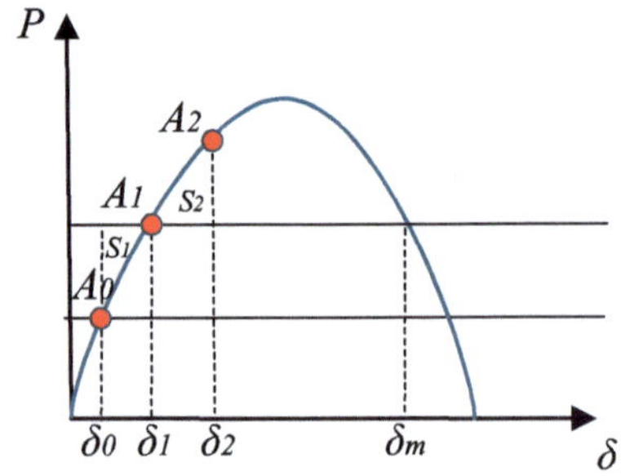

Fig. 7.5 Power-angle curve

A power-angle curve is shown in Fig. 7.5. In this figure, first the power is at A_0, and due to the system change, it increases to A_2 and then returns to A_1. Therefore, if S_1 and S_2 are equal, it shows that the acceleration power is equal to the deceleration power. It will conclude a stable case.

If the available decelerating area is sufficient to balance the accelerating area, the first swing is stable. If the accelerating area exceeds the maximum possible decelerating area, the rotor angle will continue to increase beyond the critical point, and the machine will lose synchronism.

On the other hand, during a fault, system topology will change, as shown in Fig. 7.6. Therefore, we will have three power angle curves. The sequence is:

1. Pre-fault condition: normal power-angle curve $P_e = P_{\max} \sin \delta$, operating at δ_0.
2. During fault: fault reduces P_e sharply (often to nearly zero for a fault very close to the generator), resulting in a low during-fault curve $P_e^{\text{fault}} \approx 0$.
3. Post-fault: after fault clearing (by opening the faulted line), a new power-angle curve $P_e^{\text{post}} = P_{\max} \sin \delta$ is established (usually lower than pre-fault due to loss of one circuit).

During the fault, $P_e \approx 0 < P_m$, so the rotor accelerates rapidly and δ increases from δ_0 to the clearing angle δ_c(at which the fault is cleared).

For stability:

$$S_1 = S_2 \tag{7.18}$$

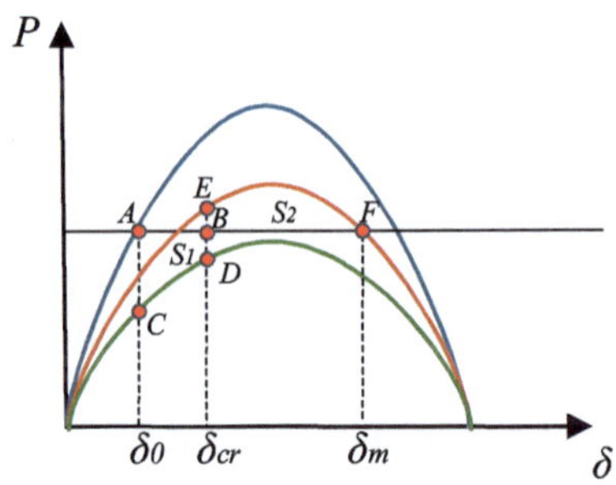

Fig. 7.6 Power-angle curve of a fault case

The critical clearing angle δ_{cr} is the value of δ_c at which A_1 just equals the maximum possible S_2(i.e., the area up to the unstable equilibrium point where $P_e^{\text{post}} = P_m$ again on the far side).

Therefore, the maximum angle that a fault can be cleared can be found by

$$\delta_{cr} = \cos^{-1}\left[\frac{P_m(\delta_m - \delta_0) + P_{\max}\cos\delta_m}{P_{\max}}\right] \tag{7.19}$$

These areas can also be investigated in the power time domain. Typically, the generators in power systems are operating at the synchronous speed with a rotor angle of δ_0 and the mechanical input power of P_m equal to the electrical output power of P_e [2]. During disturbances, the difference in power should be accounted for by the rate of change of the stored kinetic energy in the rotors of generators. It will conclude with a change in rotation speed, as follows:

$$\frac{\mathrm{H}}{\pi f}\frac{d^2\delta}{dt^2} = (P_m - P_e) \tag{7.20}$$

where H is the inertia constant, and f is the frequency. Therefore, as the rotation speed is changing, the speed deviation of the rotor can be determined as follows [3]:

$$\Delta\omega = \omega - \omega_s = \frac{d\delta}{dt} \tag{7.21}$$

where ω is the speed of the rotor during transients, ω_s is the system's frequency. By substituting (7.2) in (7.1):

$$\frac{\mathrm{H}}{\pi f} d\Delta\omega = (P_m - P_e)\mathrm{dt} \tag{7.22}$$

Then, integrating (7.3):

$$\frac{\mathrm{H}}{\pi f}\int d\Delta\omega = \int (P_m - P_e)dt \tag{7.23}$$

Two areas can be defined from (7.4), based on the fault event time t_0, and when P_e exceeds P_m time, t_1. The accelerating area, A_1, is defined as the kinetic energy released after the fault, and the decelerating area, A_2, is defined as the area where the energy is stored again in rotating masses, as shown in Fig. 7.7 and (7.5).

$$\begin{cases} A_1 = \int_0^{t_1} (P_m - P_e)dt = \frac{\mathrm{H}}{\pi f}(\Delta\omega_{t1} - \Delta\omega_{t0}) \\ A_2 = \int_{t_1}^{t_\infty} (P_m - P_e)dt = \frac{\mathrm{H}}{\pi f}(\Delta\omega_{t\infty} - \Delta\omega_{t1}) \end{cases} \tag{7.24}$$

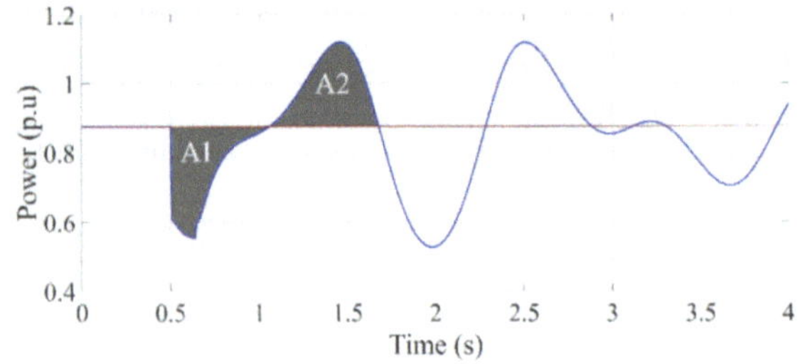

Fig. 7.7 Power-time curve of a power system

The value of $t_{\max}$ is defined as the time when, for stable conditions, the rotor's speed is equal to the synchronous speed, and for loss of synchronism conditions, it is greater than the synchronous speed.

Observing that these two terms each represent areas on the power-time curve, it could develop the so-called equal-area criterion for stability. This criterion says that stable performance requires that the accelerating area is equal to the decelerating area [4]:

$$\begin{cases} \textit{Stable Condition}\, A = A_1 + A_2 = 0 \\ \textit{Loss of Synchronism}\, A = A_1 + A_2 > 0 \end{cases} \tag{7.25}$$

7.3 Control of Active Power and Frequency

The control of active power and frequency is a fundamental requirement for maintaining the stability, security, and reliable operation of electric power systems. System frequency serves as a global indicator of the instantaneous balance between generated active power and consumer demand; even small imbalances can cause measurable deviations from the nominal frequency. When generation exceeds demand, system frequency rises, whereas when demand exceeds generation, frequency declines. If these deviations are not corrected promptly, they may lead to degraded power quality, malfunction of sensitive equipment, disconnection of generating units, or even large-scale system collapse. Modern interconnected power systems therefore employ coordinated control mechanisms that act across multiple time scales to detect imbalances, arrest frequency excursions, and restore normal operating conditions. The control layers are determined as illustrated in Fig. 7.8.

These mechanisms are organized into layered control structures that operate from local, fast-acting responses to slower, centralized supervisory controls. Immediately following a disturbance, such as a sudden load increase or generator outage, primary control acts autonomously through turbine governors to stabilize frequency by adjusting mechanical power input. Although this response limits the frequency deviation, it does not restore frequency to its nominal value. Secondary control, typically implemented through Automatic Generation Control (AGC), then intervenes to eliminate steady-state frequency errors and restore scheduled power

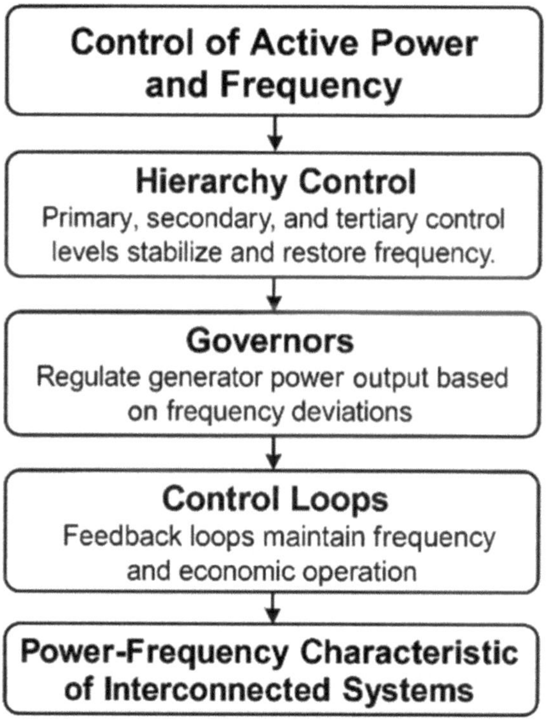

Fig. 7.8 Flow of control for active power and frequency

exchanges between interconnected areas. At a higher level, tertiary control adjusts generation dispatch and reserve allocation to optimize system operation economically while ensuring adequate contingency margins. Together, these coordinated control layers ensure that frequency disturbances are mitigated quickly, nominal operating conditions are reestablished, and sufficient reserves are maintained to withstand future disturbances, thereby supporting long-term system reliability and efficiency.

7.3.1 Hierarchy Control

Frequency control in power systems is intentionally organized as a hierarchical structure because frequency disturbances occur suddenly, while full correction and optimization require more time and coordination. Each control level operates over a different time scale and has a specific role, but all three levels work together to maintain system stability and reliability.

The primary control level is the fastest and most fundamental layer of frequency regulation. It is activated automatically and locally within a few seconds of a disturbance, such as a sudden increase in load or the tripping of a generator.

Primary control relies on the inherent speed-droop characteristics of turbine governors installed on generating units. When frequency decreases, governors increase mechanical power input; when frequency increases, they reduce it. This response is proportional to the frequency deviation and shared among participating generators. The key purpose of primary control is to arrest the initial frequency change and stabilize the system, preventing frequency from falling or rising to unsafe levels. However, because it operates with droop, primary control deliberately allows a steady-state frequency deviation and does not restore the frequency to its nominal value.

Once the system is stabilized, secondary control takes over to eliminate the remaining frequency error. Secondary control is usually implemented centrally through AGC and operates over a time span ranging from tens of seconds to several minutes. This layer continuously monitors system frequency and power flows on tie-lines between interconnected areas. Based on these measurements, AGC adjusts the active power setpoints of selected generators to restore frequency to its nominal value and to correct deviations in scheduled inter-area power exchanges. Without secondary control, frequency would remain permanently offset after disturbances, and tie-line flows between control areas would drift from agreed schedules.

The tertiary control level provides the slowest but most strategic response, acting over minutes to hours. Its primary function is not immediate frequency correction but system optimization and reserve management. Tertiary control redispatches generation by bringing additional units online, reallocating power output among generators, and transferring load from expensive or fast-acting reserves to more economical resources. This frees up secondary control capacity so the system is prepared for future disturbances. Tertiary control is often coordinated with market operations and economic dispatch mechanisms, ensuring that reliability objectives are achieved at minimum operating cost.

Together, these three hierarchical levels create a coordinated and robust frequency control structure. Primary control ensures rapid stabilization following disturbances, secondary control restores frequency and inter-area balances, and tertiary control optimizes long-term system operation and reserve availability. This layered approach allows modern interconnected power systems to respond effectively to both sudden disturbances and longer-term operational requirements while maintaining stability, security, and economic efficiency.

7.3.2 Governors

Governors are essential components in the primary frequency control of power systems because they directly regulate the mechanical input power supplied to synchronous generators. Since the electrical frequency of the power system is directly linked to the rotational speed of generator rotors, any imbalance between generated power and load demand immediately appears as a change in rotor speed. Governors continuously monitor this speed and act automatically whenever a deviation from the nominal value occurs.

When system frequency decreases due to a sudden increase in load or loss of generation, the generator rotor slows down. The governor detects this reduction in shaft speed and responds by increasing the turbine's input energy. This increase in mechanical input power counteracts the imbalance and helps stabilize the frequency. Conversely, if the frequency rises because of a sudden reduction in load, the governor reduces the turbine input to prevent excessive acceleration of the generator.

The governor's response is controlled using a droop characteristic, which establishes a proportional relationship between frequency deviation and change in mechanical power. Rather than attempting to restore the frequency to its nominal value instantly, droop control allows a controlled steady-state frequency deviation. Typical droop values of 4% mean that a small drop in frequency results in a proportional increase in power output. This characteristic is crucial in systems with multiple generators operating in parallel, as it ensures proportional load sharing among units according to their ratings.

Without droop control, generators would compete to correct frequency deviations, which could lead to instability or overloading of certain units. Droop-based governor action prevents this by distributing the required power adjustment across all participating generators, ensuring that no single unit carries an excessive share of the disturbance.

Although governors differ in their physical design depending on the type of prime mover as their fundamental purpose remains the same. All governors provide fast, local, and decentralized frequency stabilization, acting as the first line of defense against frequency deviations before slower secondary and tertiary control mechanisms take over.

7.3.3 Control Loops

Frequency regulation in electric power systems relies on a coordinated set of control loops arranged hierarchically according to their response speed and functional responsibility, allowing the system to achieve stability, accuracy, and economic efficiency simultaneously. The fastest layer is the primary control, implemented through the speed-droop loop of turbine governors, which operates locally and autonomously at each generating unit. When a disturbance causes a frequency deviation Δf, the governor adjusts the mechanical input power of the turbine according to the droop relationship:

$$\Delta P_m = -\frac{1}{R}\Delta f \tag{7.26}$$

where R is the droop constant. This proportional response acts within a few seconds to arrest the initial frequency decline or rise and stabilize the system following sudden load changes or generator outages. However, because droop control is purely proportional, it intentionally allows a steady-state frequency error, meaning the frequency does not return exactly to its nominal value after primary control action alone.

To eliminate this remaining frequency deviation, the secondary control layer, commonly referred to as load–frequency control and typically implemented via AGC, operates over a slower time scale ranging from tens of seconds to several minutes. Secondary control continuously monitors system frequency and tie-line power flows between interconnected areas and uses the Area Control Error, defined as

$$ACE = \Delta P_{tie} + B \cdot \Delta f \tag{7.27}$$

where ΔP_{tie} represents tie-line power deviations and B is the frequency bias factor. AGC processes this Area Control Error (ACE) signal using an integral control law, commonly expressed as

$$\Delta P_{ref} = -K_I \int ACE\, dt \tag{7.28}$$

The integral action ensures that the ACE is driven to zero, which implies both restoration of system frequency to its nominal value and correction of unintended inter-area power exchanges. The tertiary control layer operates on the slowest time scale, typically minutes to hours, and focuses on long-term economic optimization and reserve management rather than immediate frequency correction. Once secondary control has restored frequency and tie-line schedules, tertiary control redispatches generation resources to minimize operating costs while replenishing the primary and secondary reserves that were utilized during earlier control actions. This process is often formulated as an economic dispatch problem, in which the total generation cost $\sum_i C_i(P_i)$ is minimized subject to power balance constraints such as $\sum_i P_i = P_{load}$ and generator operating limits. Although tertiary control does not directly participate in short-term frequency dynamics, it is essential for ensuring sustainable, reliable, and cost-effective power system operation.

Together, these three control loops form a tightly coordinated frequency regulation framework in which primary control provides rapid stabilization through droop-based local action, secondary control restores frequency and inter-area balance using ACE-driven integral control, and tertiary control ensures economical operation and reserve availability. Proper coordination and tuning of these layers are critical, as poor interaction can result in oscillations, instability, or inefficient utilization of generation resources, whereas well-designed coordination ensures both dynamic stability and long-term reliability of interconnected power systems.

7.3.4 Power–Frequency Characteristic of Interconnected Systems

The power-frequency characteristic describes the overall interconnected system responds to changes in active power. It depends on the combined inertia of all synchronous machines, the aggregate droop settings of participating generators, and the frequency sensitivity of electrical loads. In multi-area systems, interconnected behavior is quantified through the ACE, which combines deviations in tie-line power flow with frequency deviation scaled by a frequency-bias factor. Secondary control continuously acts to drive ACE toward zero, thereby maintaining both nominal frequency and scheduled inter-area power exchanges. Primary control ensures proportional load sharing among areas following disturbances, while secondary control handles long-term restoration. As modern grids integrate higher levels of inverter-based renewable energy, overall inertia decreases, causing frequency to change more rapidly during disturbances and increasing the need for fast frequency response solutions such as synthetic inertia from wind turbines, battery storage, and other advanced power-electronic sources.

7.4 Control of Reactive Power and Voltage

It is an important topic in power systems because it ensures that electricity is delivered at safe and stable voltage levels. Voltage in an electrical network depends largely on reactive power, which is required to maintain the electric and magnetic fields in power lines, transformers, and motors. Although reactive power does not do useful work like active power, it plays a crucial supporting role by allowing active power to be transmitted effectively through the system.

If reactive power is not properly managed, voltage problems can occur. Too little reactive power causes voltage to drop, which can lead to poor performance of electrical equipment or even power outages. On the other hand, too much reactive power can raise voltage above safe limits and damage system components. To prevent these problems, power systems use a combination of generators, capacitors, reactors, and modern power-electronic devices to control reactive power and regulate voltage. By coordinating these methods, utilities maintain voltage within acceptable limits, improve power quality, reduce losses, and ensure reliable operation of the electrical grid.

7.4.1 Generation and Absorption of Reactive Power

In a power system, reactive power is constantly being produced and used by different parts of the network. Sources of electricity, like generators, as well as equipment such as transmission lines, transformers, and special devices, all play a role in handling reactive power.

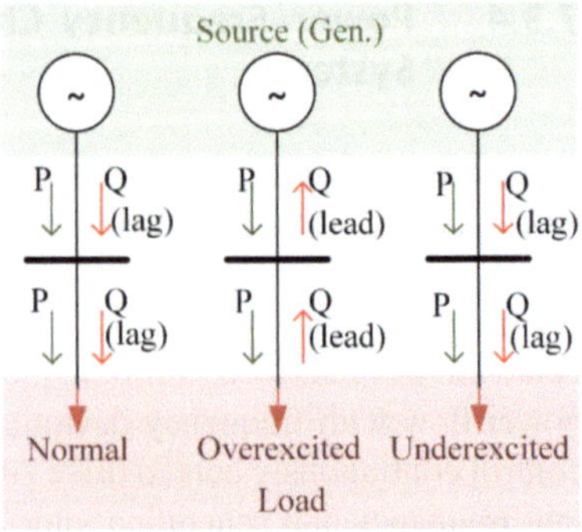

Fig. 7.9 Normal, underexcited and overexcited loads

Generators are one of the main and most important sources of reactive power because they can control production or absorption. This control is done by adjusting the generators excitation, which is like controlling the strength of its magnetic field. When a generator is over-excited, it produces extra reactive power, which helps increase the voltage near that generator. When it is under-excited, it absorbs reactive power, which causes the voltage to decrease as shown in the Fig. 7.9. This makes generators very useful for controlling voltage levels in the system.

Transmission lines also affect reactive power. Long, high-voltage lines can naturally produce reactive power because of their electrical properties, especially their capacitance. However, when lines carry heavy currents, they tend to absorb reactive power due to their inductive nature. This means that the same line can either supply and also consume reactive power depending on its operational properties. Most electrical loads, such as motors, pumps, and industrial machinery, usually consume reactive power. These devices need reactive power to create magnetic fields so they can operate properly. The reactive power they use must be supplied by nearby generators or compensation devices to keep voltage at a safe level.

It is very important to keep a balance between reactive power sources and reactive power users. Unlike active power, reactive power cannot travel efficiently over long distances, so it needs to be supplied close to where it is used. If this balance is not maintained, voltage problems can occur. Therefore, reactive power must be carefully managed locally within different parts of the power system to ensure stable and reliable operation.

7.4.2 Methods of Voltage Control

Voltage control means keeping the voltage at the right level so electrical devices work safely and properly. One way to control voltage is by using transformers. Some transformers can change their settings to raise or lower the voltage when needed. This helps keep the voltage steady as electricity moves through the network. Generators also help control voltage. By adjusting their operation, generators can supply or absorb reactive power, which directly affects voltage. Supplying reactive power increases voltage, while absorbing it decreases voltage.

However, for the power systems to apply as using we use capacitors and reactors. Capacitors aids in increase voltage by supplying reactive power, and reactors reduce voltage by absorbing reactive power. These devices are simple and are often used during normal operation. In modern power systems, fast electronic devices are also used. These devices react very quickly when voltage suddenly changes, such as during a fault or a sudden increase in load. By using all these methods together, the power system keeps voltage stable, protects equipment, and ensures electricity is delivered safely and reliably.

7.4.2.1 Static Shunt Capacitors

Static shunt capacitors are among the simplest and most widely used devices for voltage control and reactive power compensation in power systems. They are connected in parallel (shunt) to the network, typically at substations or along distribution feeders, and their main purpose is to supply reactive power near the load. By doing so, shunt capacitors help raise local voltage levels, reduce the amount of reactive power that must flow through transmission lines, and lower system losses, making power delivery more efficient.

The reactive power supplied by a shunt capacitor depends on the system voltage and is given by the relationship

$$Q_C = \frac{V^2}{X_C} \tag{7.29}$$

where Q_C is the reactive power supplied by the capacitor, V is the voltage at the connection point, and X_C is the capacitive reactance. This equation shows that the reactive power output of a shunt capacitor is proportional to the square of the voltage. As a result, shunt capacitors are most effective when the voltage is close to its nominal value. At normal operating conditions, they provide steady and reliable reactive power support, improving voltage profiles and reducing transmission losses.

The voltage drops along the feeder can be represented as:

$$V_d = I_r R + I_x X_L \tag{7.30}$$

where, the total resistance fed to the circuit is represented by R, whereas total reactance (inductive) is denoted by X_L. The current impacting the real power is denoted by I_r whereas the reactive power is impacted by I_x. This behavior can also be understood using a phasor diagram denoted in Fig. 7.10. In an AC system, a shunt capacitor draws a current that leads the voltage by 90º. This leading current corresponds to leading reactive power, which offsets the lagging reactive power absorbed by inductive loads such as motors and transformers. When the capacitor injects this leading reactive power into the system, the overall reactive demand decreases, causing the voltage magnitude to increase. In phasor terms, the

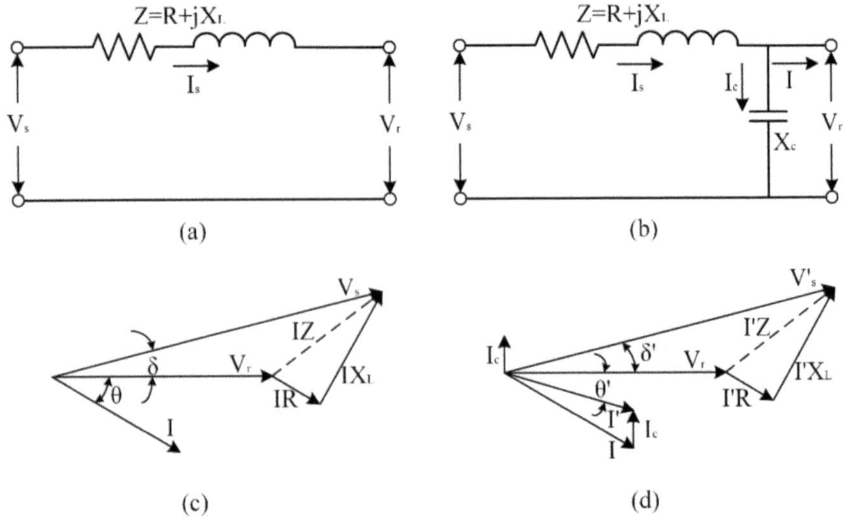

Fig. 7.10 Phaser diagram **a** feeder without the shunt capacitor, **b** feeder with the shunt capacitor, **c** regular phasor, **d** phasor for shunt capacitor

capacitive current phasor lies ahead of the voltage phasor, directly illustrating how capacitors support voltage.

7.4.2.2 Static Series Capacitors

Static series capacitors are devices that are installed directly in series with transmission lines to improve voltage performance and increase power-transfer capability, especially over long distances. Transmission lines are naturally inductive, meaning they oppose the flow of current and cause voltage drops when power is transmitted. By inserting capacitors in series with the line, part of this inductive effect is cancelled, allowing power to flow more easily and improving overall system performance.

From an electrical point of view, the series capacitor reduces the total series reactance of the transmission line. If the original line reactance is X_L and the capacitor reactance is X_C, the effective reactance becomes

$$X_{\text{eff}} = X_L - X_C \tag{7.31}$$

Because the maximum power transmitted over a line is given by

$$P_{\max} = \frac{V_1 V_2}{X_{\text{eff}}} \sin \delta \tag{7.32}$$

reducing X_{eff} directly increases the maximum power-transfer capability of the line. This is one of the main reasons series capacitors are used in long, heavily loaded transmission corridors.

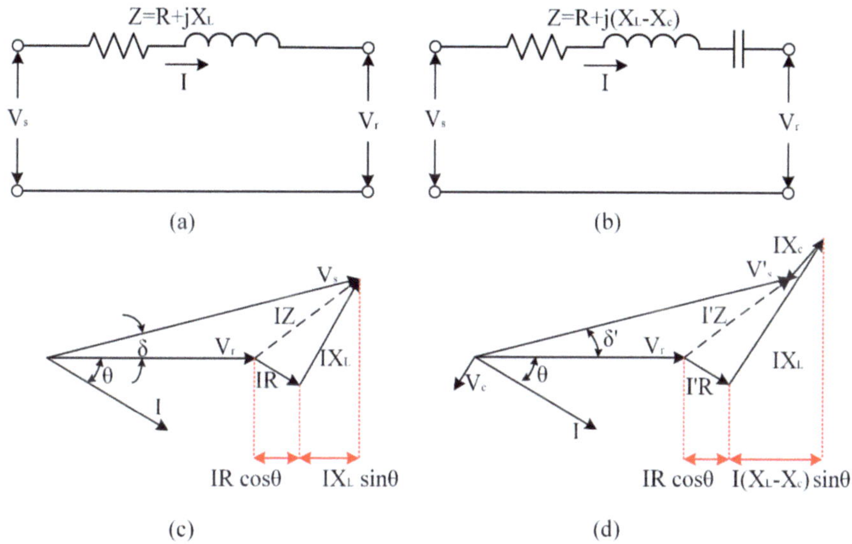

Fig. 7.11 Phaser diagram **a** feeder without the series capacitor, **b** feeder with the series capacitor, **c** regular phasor, **d** phasor for series capacitor

In addition to increasing power transfer, static series capacitors also help improve voltage regulation along the transmission line. During heavy loading conditions, inductive reactance causes large voltage drops. By reducing this reactance, series capacitors decrease the voltage drop, allowing the receiving-end voltage to remain closer to its nominal value. This makes power delivery more stable and efficient, particularly in extra-high-voltage (EHV) systems.

The voltage drops along the feeder can be represented as:

$$V_d = I_r R \cos\theta + I_x X_L \sin\theta \tag{7.33}$$

where, the total resistance fed to the circuit is represented by R, whereas total reactance (inductive) is denoted by X_L. The reactive end power factor is impacted by $\cos\theta$. The operation of series capacitors can also be understood using a phasor explanation in Fig. 7.11. In a transmission line, the line current typically lags the voltage because of inductive reactance. When a series capacitor is added, it produces a voltage that lags the current by 90 degrees, opposite to the inductive voltage drop. In phasor terms, the capacitive voltage cancels part of the inductive voltage drop, reducing the net voltage drop across the line. This phasor cancellation explains why series capacitors improve voltage levels and increase real-power transfer.

7.4.2.3 Synchronous Compensators

Synchronous compensators, also called synchronous condensers, are rotating machines that are connected to the power system without any mechanical load,

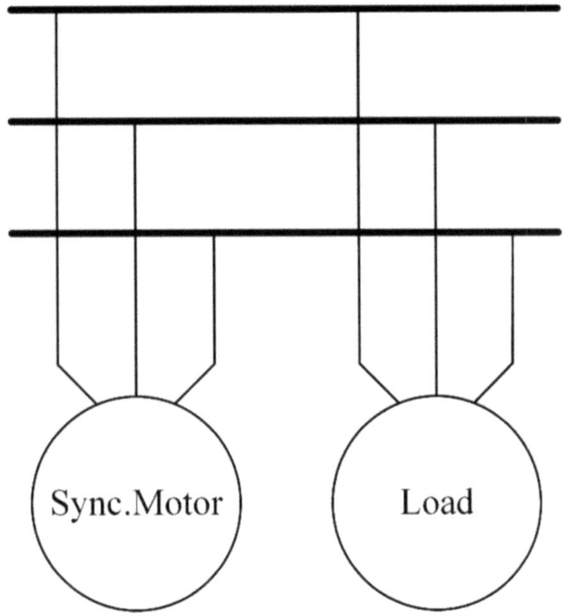

Fig. 7.12 Power factor correction by coordinating synchronous machine with load

as shown in Fig. 7.12. Unlike generators that produce electricity, their only purpose is to control reactive power and support voltage. They operate in the same way as synchronous motors or generators, but without delivering real power.

The reactive power output of a synchronous compensator is controlled by its excitation system. During the over-excited operation, the machine supplies reactive power to the system and raises the voltage. Whereas during the under-excited operation, it absorbs reactive power and lowers the voltage. Thus, a synchronous compensator can smoothly shift between reactive-power generation and absorption depending on system needs.

This behaviour can be expressed conceptually as

$$Q \propto (E - V) \tag{7.34}$$

where E is the internal generated voltage controlled by excitation, and V is the system voltage. If $E > V$, the compensator supplies reactive power; if $E < V$, it absorbs reactive power.

As per the phaser illustrated in the Fig. 7.13, the synchronous compensator changes the phase relationship between voltage and current by adjusting its excitation. Over-excited operation produces a leading current, while under-excited operation produces a lagging current. This flexibility makes synchronous compensators very effective for voltage control.

Unlike static devices, synchronous compensators also contribute rotational inertia and short-circuit strength to the system. This makes them especially valuable in weak grids, where voltage can change rapidly during faults or switching events.

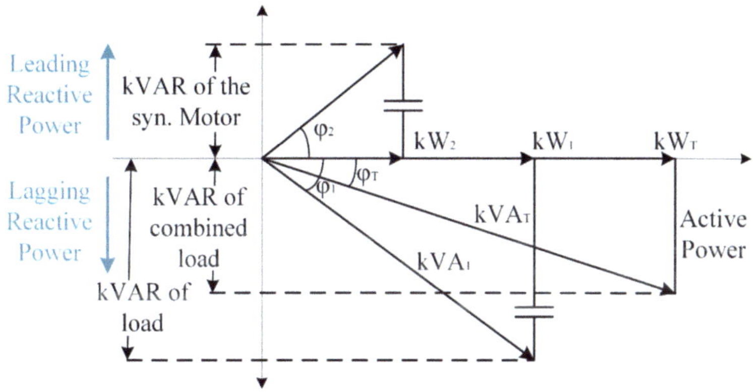

Fig. 7.13 Phasor diagram related to synchronous compensators

They respond quickly to voltage disturbances and provide strong dynamic voltage support, helping stabilize the system during transient conditions.

7.4.2.4 Static Var Compensators (SVCs) and STATCOMs

Static Var Compensators (SVCs) and Static Synchronous Compensators (STATCOMs) are power-electronic devices designed to provide fast, accurate, and continuous reactive power control. They are widely used to maintain voltage stability, reduce flicker, and improve system performance under rapidly changing conditions.

An SVC uses combinations of thyristor-controlled reactors (TCRs) and thyristor-switched capacitors (TSCs) to adjust reactive power as represented in Fig. 7.14. By controlling the firing angle of the thyristors, the SVC can smoothly vary the reactive power it absorbs or supplies. The reactive power output of an SVC depends on the system voltage and can be approximated as

$$Q_{\mathrm{SVC}} = V^2 \cdot B_{\mathrm{var}} \tag{7.35}$$

where B_{var} is the controllable susceptance. This means the effectiveness of an SVC decreases during deep voltage dips.

A STATCOM is a more advanced device based on a voltage-source converter (VSC). Instead of switching capacitors and reactors, a STATCOM generates an AC voltage internally and exchanges reactive power with the grid by adjusting the magnitude of this voltage as illustrated in Fig. 7.15. If the converter voltage is higher than the system voltage, reactive power flows to the system.

As a result, STATCOMs perform much better during severe voltage drops, making them ideal for weak grids, renewable-energy-dominated systems, and applications requiring fast dynamic voltage support. They also contribute to improved fault-ride-through capability and system stability.

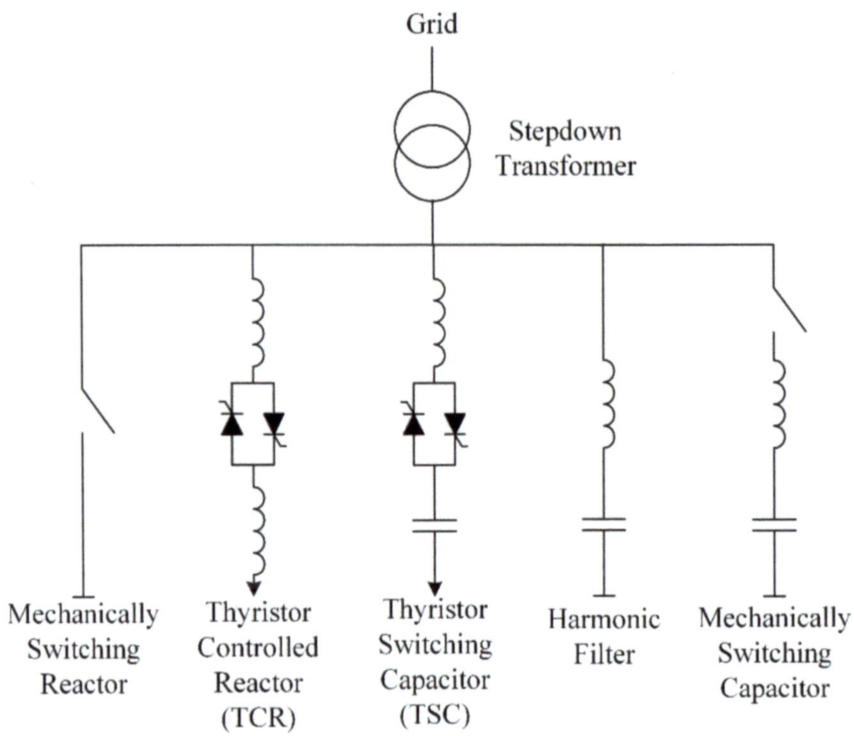

Fig. 7.14 Single line diagram with typical SVC configuration

Fig. 7.15 Single line diagram with STATCOM configuration

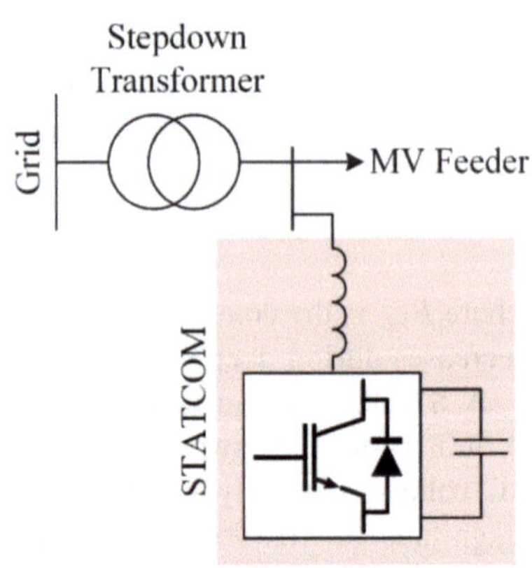

7.5 Stability and Control Study in PowerFactory

In this section, as an example, a single generator with a type of 46 MVA is connected to two loads, each 20 MW, as shown in Fig. 7.16. Initially, one of them is not connected, and as a large load switching disturbance, it will connect after 2 s. Then, the voltage of the generator bus drops, and frequency increases rapidly, as shown in Fig. 7.17. It shows instability in this system. To solve this problem, a governor and an excitation system should be connected to the generator.

7.5.1 Frequency Control

By right-clicking on the generator and selecting Network Models, Governor and Turbine (gov), and New, a governor system can be installed at the generator, as shown in Fig. 7.18.

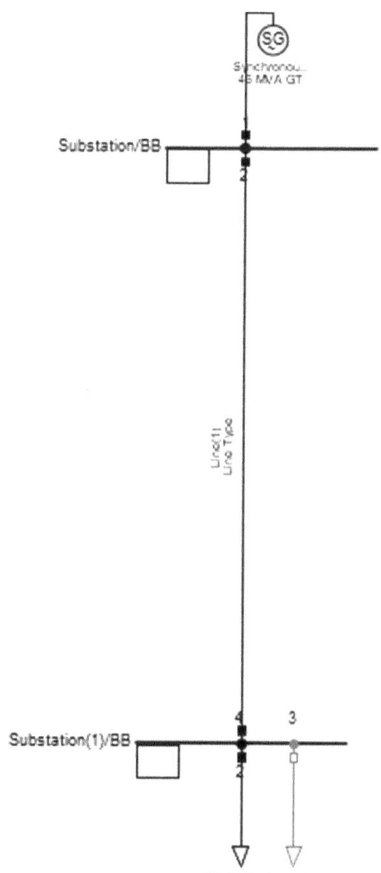

Fig. 7.16 Single machine system example in PowerFactory

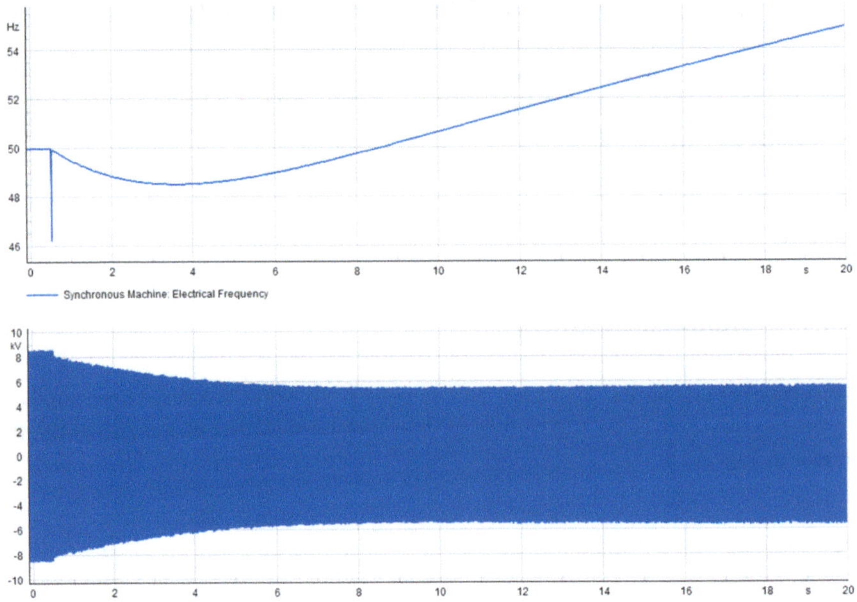

Fig. 7.17 Frequency and voltage for large load change without controllers

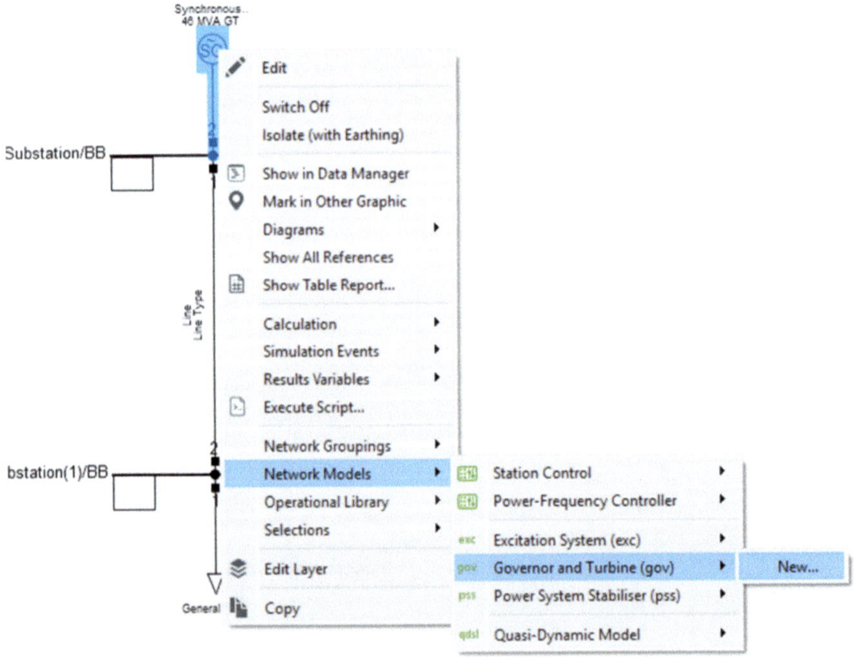

Fig. 7.18 Installing the governor in PowerFactory

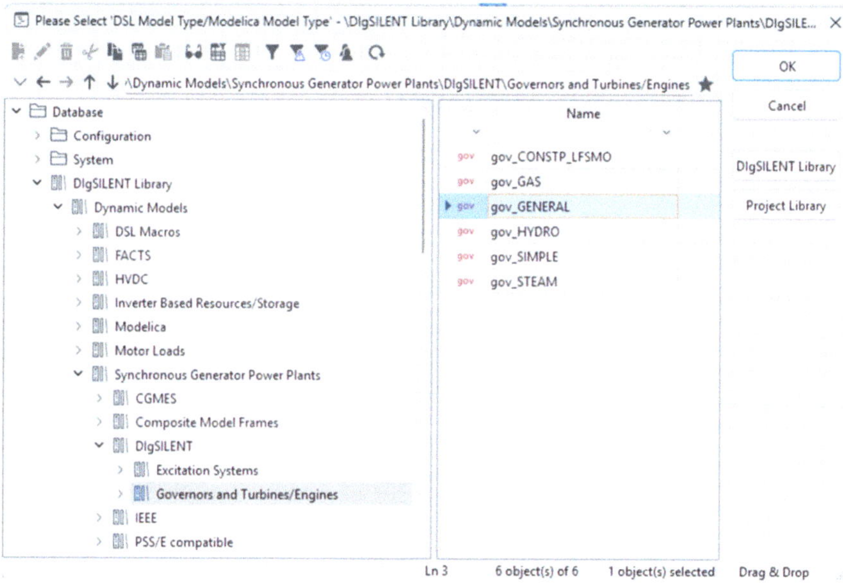

Fig. 7.19 Governors in PowerFactory

Then, in the new window, by going to the library shown in Fig. 7.19, different governors are predefined by PowerFactory, and in this example, the gov_GENERAL is selected, with parameters shown in Fig. 7.20.

7.5.2 Voltage Control

By right-clicking on the generator, selecting Network Models, and then Excitation system (exc), and New, an excitation system can be installed at the generator, as shown in Fig. 7.21. Then, in the new window, by going to the library shown in Fig. 7.22, different excitation systems are predefined by PowerFactory, and in this example, the exc_SIMPLE is selected.

After installation of the governor and excitation systems, and running the simulation again, the frequency and voltage of the governor are under the normal operating in stable mode, as shown in Fig. 7.23.

7.6 Summary

This chapter has presented the fundamental concepts of power system stability and control, which are essential for ensuring reliable, secure, and continuous operation of modern electric grids. It began by classifying stability into voltage

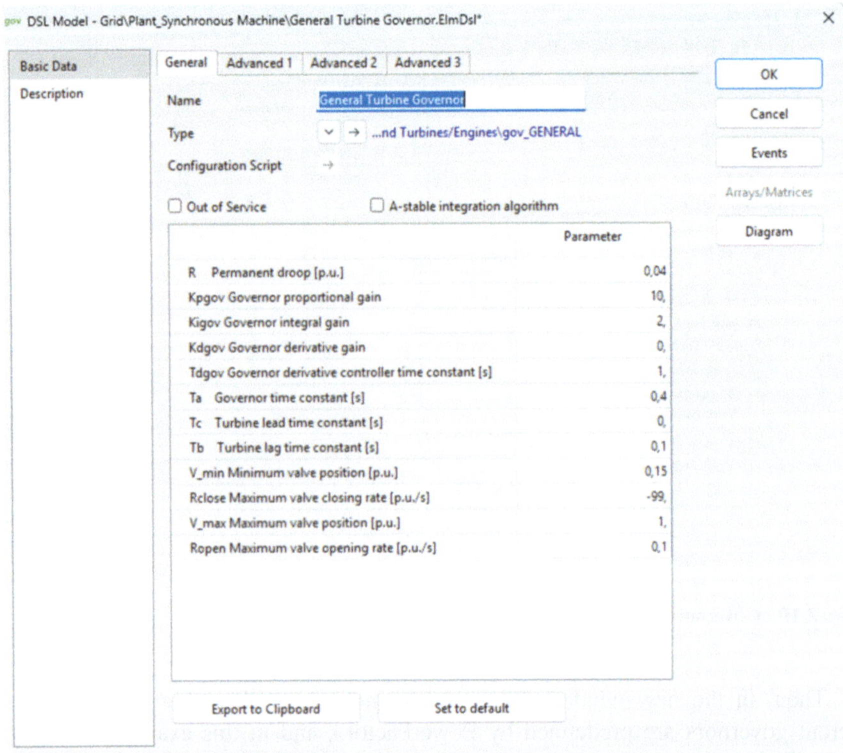

Fig. 7.20 Governor parameters in PowerFactory

stability and frequency stability, highlighting how disturbances such as short circuits, large load changes, or generator outages can lead to unacceptable voltage drops, frequency excursions, or loss of synchronism. The dynamics of synchronous machines were then examined through the swing equation, the power-angle characteristic, and the single-machine infinite-bus model, providing the theoretical basis for understanding rotor-angle behavior under both small and large disturbances.

Steady-state stability was discussed using the synchronizing power coefficient, while transient stability was analyzed with the equal-area criterion in both the power-angle and power-time domains. These analytical tools enable engineers to determine critical clearing angles and times without full numerical integration of the swing equation. The chapter then moved to practical control strategies. Active-power and frequency control were structured hierarchically: primary (governor droop) for rapid stabilization, secondary (AGC via area control error) for frequency restoration and tie-line scheduling, and tertiary for economic dispatch and reserve management. The power–frequency characteristic of interconnected systems and the impact of reduced system inertia due to inverter-based resources were emphasized.

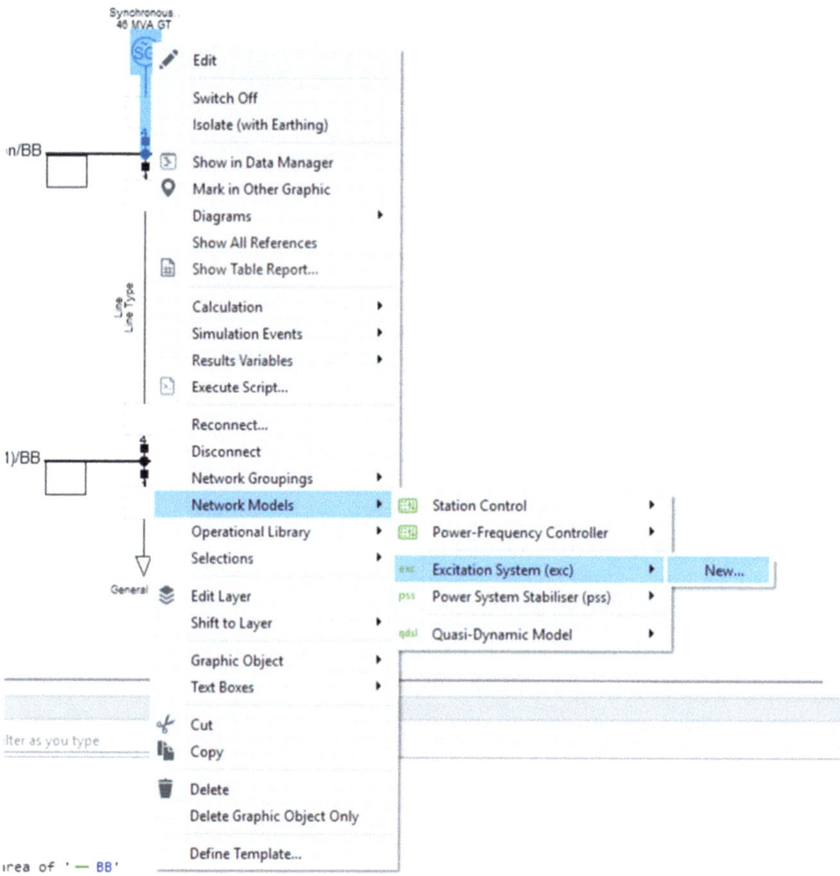

Fig. 7.21 Excitation in PowerFactory

Reactive-power and voltage control were addressed next, covering the generation and absorption of reactive power by generators, transmission lines, and loads. Voltage regulation methods were presented in detail, including static shunt and series capacitors, synchronous compensators, SVCs, and STATCOMs, with phasor diagrams and governing equations illustrating their operating principles and relative advantages. Finally, a complete step-by-step PowerFactory tutorial demonstrated the implementation of governor and excitation systems on a single-machine example, showing how frequency and voltage remain stable after a large load-switching disturbance when proper controllers are applied.

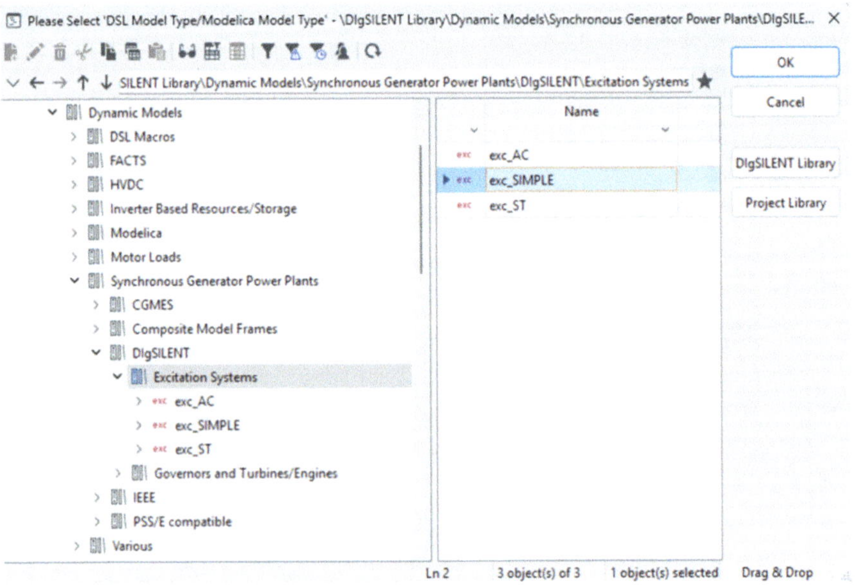

Fig. 7.22 Excitations in PowerFactory

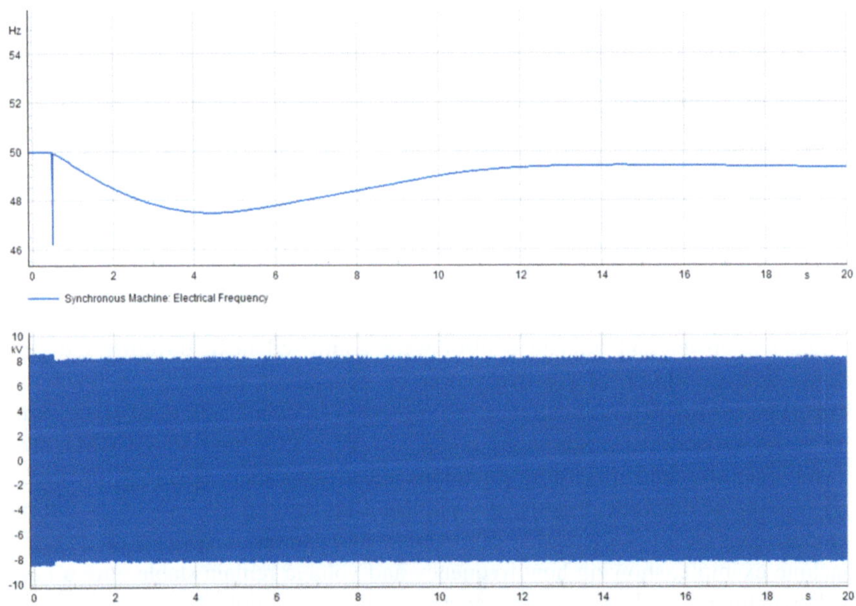

Fig. 7.23 Frequency and voltage for large load change with controllers

7.7 Problems and Questions

1. A synchronous generator is connected to an infinite bus through a lossless line with total reactance $X = 0.8$ pu. The generator internal voltage is $E = 1.2$ pu and the infinite-bus voltage is $V = 1.0$ pu. The mechanical power input is constant at $P_m = 0.9$ pu. (a) Calculate the initial rotor angle δ_0. (b) Using the equal-area criterion, determine the critical clearing angle δ_{cr} if a three-phase fault reduces electrical power to zero during the fault. (c) If the fault is cleared at $\delta = 60°$, determine whether the system remains stable and calculate the maximum rotor angle δ_m.
2. A generator operates with a 5% droop governor and supplies 0.8 pu power at nominal frequency. A sudden load increase of 0.3 pu occurs on the system. (a) Calculate the steady-state frequency deviation after primary control action. (b) If the system frequency bias factor $B = 1.2$ pu/Hz and the area control error must be driven to zero, explain the role of secondary control in restoring frequency. (c) Sketch the expected frequency response curve showing primary and secondary control contributions.

References

1. Rezkalla M, Pertl M, Marinelli M (2018) Electric power system inertia: requirements, challenges and solutions. Electr Eng 100(4):2677–2693
2. Ramar S, Kuruseelan S (2013) Power system analysis. PHI Learning Pvt. Ltd.
3. Paudyal S, Ramakrishna G, Sachdev MS (2010) Application of equal area criterion conditions in the time domain for out-of-step protection. IEEE Trans Power Deliv 25(2):600–609
4. Bayati N, Ebel T (2025) Application of supercapacitors to extend the critical clearing time by time domain equal area criterion. In: 2025 IEEE 66th international scientific conference on power and electrical engineering of Riga Technical University (RTUCON) 2025 Oct 23. IEEE, pp 1–6

… and Questions

MIX
Papier aus verantwortungsvollen Quellen
Paper from responsible sources
FSC® C105338

If you have any concerns about our products,
you can contact us on
ProductSafety@springernature.com

In case Publisher is established outside the EU,
the EU authorized representative is:
Springer Nature Customer Service Center GmbH
Europaplatz 3, 69115 Heidelberg, Germany

Printed by Libri Plureos GmbH
in Hamburg, Germany